Think Like a
Mathematician

Also by Junaid Mubeen

Mathematical Intelligence:
What We Have That Machines Don't

Think Like a Mathematician

SIMPLE TOOLS
FOR COMPLEX
EVERYDAY PROBLEMS

Junaid Mubeen

PEGASUS BOOKS

NEW YORK LONDON

THINK LIKE A MATHEMATICIAN

Pegasus Books, Ltd.
148 West 37th Street, 13th Floor
New York, NY 10018

Copyright © 2026 by Junaid Mubeen

First Pegasus Books cloth edition February 2026

Diagrams used with permission: p. 89, Booyabazooka, English Wikipedia;
p. 145, Flickr; p.191, Kazukiokumura, Creative Commons;
p. 227, NerdBoy 1392, Creative Commons; p. 282, Woraput, iStock.

ISBN: 979-8-89710-046-0

10 9 8 7 6 5 4 3 2 1

Printed in the United States of America
Distributed by Simon & Schuster
www.pegasusbooks.com

For Elias,
the coolness of my eyes

Contents

Introduction

Here's to maths, and its surprising usefulness

In the summer of 2011, during my final weeks as a maths student at Oxford, I chanced upon an unusual conversation. The features of a maths common room are unmistakable: equations scribbled onto whiteboard tables, animated discussion of highly technical concepts, copious consumption of coffee, a hotly contested game of chess.

As I worked my way through the communal biscuits, my ears pricked up as two of my fellow students shared their latest life updates. What initially caught my attention was the contrast between their two speaking styles: the measured Brit on the one hand, an American who exuded scarcely believable levels of enthusiasm with every syllable on the other. But what held my attention was their topic of conversation.

Brit: I'm spending this weekend with my girlfriend.

American: Wow, sounds serious. Do you think she's the one?

Brit: Well ... I believe in $n + 1$.

American: $n + 1$ – that's deep, man. But what does it mean?

Brit: Well, I love her today, and I believe that if I love her at any point in time, then I'll love her at the next point in time. By the principle of mathematical induction, the logical conclusion is that my love for her will endure forever. So yes, in that sense, I guess she is the proverbial *one* ...

I may have paraphrased, but only slightly. Our enthused American was in raptures, having been granted the secret to everlasting love, which apparently resides in a mathematical concept known to every maths undergraduate.

The encounter took place during my transition from the ivory towers of academia to the 'real world', where I would soon have to convince employers of my subject's relevance. As bemusing as the conversation was, it offered a clue on how to answer that oft-asked question concerning maths: *when will we ever need this?*

Mathematicians often find themselves on the defensive when society demands they explain how their work benefits others. When they do serve up a response, it usually takes one of three forms. The first, which is the most direct, draws on the practical benefits of maths in tackling real-world problems. Mastering percentages and fractions helps us to spot bargains (and rip-offs) at the supermarket. Engineers use trigonometry to design sturdy structures. A grasp of statistics equips us to analyse headline-grabbing claims. Probability gives us an edge in the stock market and helps us forecast election outcomes. The benefits of so-called 'applied maths' go on and, for many people, they provide the only justification maths needs.

But try telling that to so-called 'pure' mathematicians, who spend most of their time working on abstract concepts that are decidedly removed from real-world consideration. They liken maths to an art form and speak of the subject's aesthetic qualities.[1] They seek beautiful patterns and elegant arguments, with no regard for utility. Pure mathematicians often take pride in the apparent uselessness of their work, even deriding the supposed need for their subject to bring practical benefits. 'Here's to pure mathematics', starts one toast, 'may it never be of use to anyone.'[2] It is an elitist view – and one that I must confess to being endeared by as a student. It was only when I left academia, and my salary suddenly depended on the applicability of my skills, that I felt an impetus to justify the subject in more concrete terms.

The third response bridges these views by suggesting that even the most abstract and arcane mathematical ideas have a knack of finding their way to some application or another. Prime numbers, fractals and imaginary numbers all originated in the playground of pure mathematics but now underpin the world in which we live, from internet security (primes) to the design of cities (fractals) and the transmission of radio waves (imaginary numbers). This is what the American educator Abraham Flexner meant by the 'usefulness of useless knowledge' in his 1939 essay of the same title, when he argued that many of humankind's greatest scientific breakthroughs owe a huge debt to scientific inquiry that may initially be dismissed as pointless.[3]

My own answer to the question *'Why maths?'* is a variation on this theme. I have come to appreciate the subject as a collection of portable thinking tools that enrich the way we see the world, even when we do not expect them to. Mathematicians spend so much of their time in the abstract realm that the objects they study routinely seep into their ways of thinking. Our wise Brit, after all, did not learn the principle of induction with love in mind. Mathematicians study these concepts because they bolster their efforts to solve abstract problems – yet the more they mull over ideas in the mathematical universe, the more those ideas imprint themselves in their minds and influence their thoughts and behaviour.

Spend enough time with a mathematician and you'll uncover the telltale signs. They will speak with agonisingly precise vocabulary that reflects the rigour of their subject. Their conversations – even those that are not about mathematics – are filled with words like 'orthogonal' (irrelevant) and 'modulo' (except for). In their approach to everyday problems, mathematicians will naturally, even subconsciously, resort to the models they have acquired and embedded through years of study.

The usefulness of abstract maths should not come as a complete surprise. Many concepts derive in the first instance from a real-world context. Take the example of dimensionality. We are attuned to perceive the world in up to three physical dimensions and have a natural sense of what each one represents. Mathematics, as we shall see, gives us a simple and precise framework to describe objects in one, two or three dimensions. But before we know it, this framework has us making sense of four, five, even infinite-dimensional spaces. It's heady stuff, abstraction to its core. But the pay-off, aside from the interesting maths that arises, is that it gives us a more expansive way of thinking about space and, as Chapter 8 explores, it extends our understanding of various real-world phenomena beyond the usual three dimensions.

This book is a catalogue of mathematically inspired mental models – concepts that enrich our worldview and help us approach everyday situations. With a few exceptions, I have steered clear of more obvious examples. Well-trodden topics from statistics and related fields receive little attention here; there are plenty of excellent books that show us how to make sense of data. This book is focused on more abstract mathematical ideas that you probably will not have encountered before – or if you have, will show them in a new and unexpected light.

The majority of the mental models in this book are sourced from undergraduate courses. We will shed them of their technical baggage and focus only on the big-picture ideas they represent. If you are currently studying for a maths-related degree, consider them a preview of how abstraction will inevitably shape your understanding of the world. If you are a self-proclaimed mathophobe, do not be deterred – these models are eminently graspable, regardless of your prior experiences with the subject. Mathematicians benefit enormously from the thinking tools of their trade, and it turns out that everyone else can too with surprisingly little effort.

...

No mental model should be above reproach and you might already have taken issue with the Calm Brit's account of falling in love. At the time of that exchange in the common room, I was a bachelor in my mid-twenties, with no immediate prospect of love. It's fair to say

I was intrigued; maybe the same path lay ahead for me – an application of mathematical induction that would land me a life partner. But I was also sceptical. Were love reducible to logic, it would save the world from so much heartbreak. Mathematicians would have romantic partners queuing up (which, for the avoidance of doubt, they do not). Hollywood would rewrite its scripts, the mysteries of romance giving way to cold, calculated reasoning. Love, in all its maddening complexity, defies any serious attempt at mathematical description.

Attempts to 'mathematise' the world can undoubtedly go too far. There is a wonderful scene in the film *Dead Poets Society* when schoolteacher John Keating, played by Robin Williams, asks a student to read a passage from a textbook that references the Pritchard Scale of Understanding Poetry, in which the supposed 'greatness' of a poem can be quantified with the formula

$$\text{Greatness} = \text{Perfection} \times \text{Importance}$$
$$(\text{or } G = P \times I, \text{ in shorthand})$$

A sonnet by Byron, according to the passage, scores high for I and moderately for P, whereas one by Shakespeare scores high on both. By placing P on a horizontal axis and I on a vertical axis, we can visualise a poem's overall quality in terms of the area under the rectangular bar it forms, with Shakespeare occupying a greater area than Byron:

Importance (I)

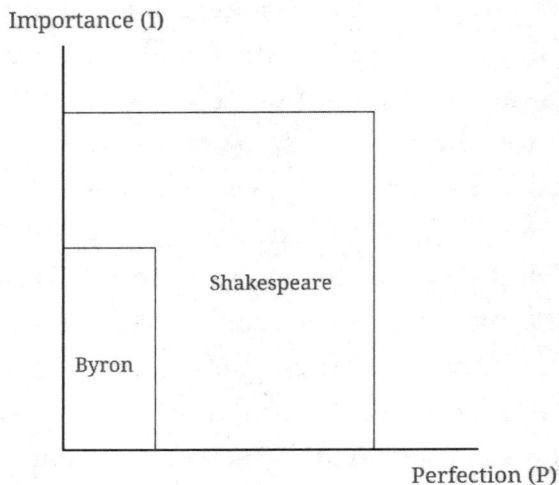

Shakespeare

Byron

Perfection (P)

As his students diligently copy out the passage, Mr Keating gives a one-word verdict on the approach, 'excrement', instructing the students to tear the page from their books. The sciences are necessary to sustain life, Mr Keating explains, but pursuits like poetry are what we stay alive for.

Taking my cue from Mr Keating, I will declare outright that this book does not contain a prescription for 'the good life', whatever that means. That's not the point of mental models. Rather, they should illuminate our ways of thinking – bad as well as good. The Calm Brit's choice of model – induction – rests on the questionable assumption that stable patterns will persist in the future. The assumption often holds up in the realm of numbers, but it's an example of abstraction gone too far: the real world is a little too messy, our lives a little too unpredictable, than his argument implies. We'll return to induction in Chapter 6, by which point we will have encountered a host of other models that go some way to explaining where the Calm Brit falls short. None will quite crack the code for love, but they might offer a new perspective on how relationships develop. The reason induction has made it into this book is precisely because it exposes a flaw in our everyday thinking – not just in matters of love but in how we cling to the orderliness of our daily routines and refuse to countenance how they might be uprooted at any moment.

The models in this book are simply lenses through which to see the world. A few you will adopt right away because they bring clarity and depth to your view of a particular situation. Others will help you to understand how human thinking can go awry. Some will feel like folk wisdom because a lot of mathematical concepts are just formalisations of things we know to be blindingly obvious, but that we sometimes need reminding of in a world awash with misinformation and bad advice. There are reflections on productivity and parenting, on politics and sport, on relationships and on the idiosyncrasies that fill our lives. Some of the models are intended to be serious, others light-hearted. All of them are a matter of perspective. At the very least, if you ever find your-self surrounded by caffeine-fuelled mathematicians, you might just find yourself able to make sense of their musings.

So here's to pure maths – may we all benefit from its usefulness. And here's to you for daring to think like a mathematician.

1

The Continuum

The infinitesimal detail of the everyday

Several years ago, I went to a doctor seeking help for a chronic stomach condition that causes bouts of severe pain. I was asked to mark my discomfort on a five-point chart of emojis. The leftmost emoji was a picture of serenity, very different from the anguished-looking emoji on the far right, with the ones in between representing moderate levels of discomfort. I was dumbstruck by how, for all the marvels of modern medicine, my pain assessment boiled down to this most blunt of instruments.

You might have felt similarly restricted when asked to rate your hotel stay, or your Uber ride or, dare I suggest, the book you are reading. Five-star rating systems are very much the norm, but your choices are limited to a handful of whole numbers; values in between are strictly off limits. While responses are averaged into a more precise value (usually given to one or two decimal places), for the individual rater the options can feel inadequate. The ride was smooth but not perfect – four stars would be too harsh, five stars too generous. What's more, with the pressure exerted on gig economy workers – where, for instance, Uber drivers are required to maintain ratings upwards of 4.6 – the rating scale becomes distorted. A rating of five stars is now expected, and even a four-star rating is viewed as an indictment of one's work and afflicts reputational damage. Feedback is reduced to a perfection-or-bust logic.

Social media content is often liable to the same issue – the 'heart' icon on sites like Instagram and TikTok, for instance, enforces the most dichotomous of responses to each post (where not clicking the heart is taken as disapproval or disinterest), allowing no room for an individual to register a nuanced opinion. The primary purpose of these buttons (beyond serving as a measurable unit of engagement for marketers) is not for us to reflect on the content in any meaningful way, but to help shape our algorithmically optimised feeds and to present us with more of the same. Adding to the problem, the comments section typically reinforces the most commonly held views in either direction. This problem is magnified on Reddit, a platform that is ostensibly designed for discussion, where users have the option of either upvoting or downvoting a post, as well as its comments. This system, which at first appears equitable, penalises unpopular opinions because downvoted comments are quickly concealed from view and collapsed into an ignominious section at the bottom of the post. Genuine debate is rendered almost impossible.

Discrete rating scales persist in spite of a readily available alternative: the sliding scale. It's a startingly simple idea: instead of deciding whether you like or dislike a post, the question now becomes one of degrees, the *extent* to which it swayed your opinion in either direction. Instead of pinning your Uber driver to one of five ratings you would have license to place them anywhere on the scale from 1 to 5, whole number or otherwise. And fluid concepts such as consciousness could be viewed not in terms of binary on/off labels but an entire progression of states between those two extremes.

The sliding scale gets overlooked because it requires more mental effort on the part of the rater. Like many facets of online decision-making, feedback systems are designed to reduce our cognitive burden and to mindlessly push us towards quick and simple actions. Limiting our choices to predefined marks on a scale is akin to thinking the rainbow is made up of just the seven colours of ROYGBIV we learned in school. Those colours represent specific wavelengths (in increasing order) but we could just as well reference ten colours, or a hundred, or indeed any number between the two extremes of red and violet. Isaac Newton, whose experiments led to the discovery of the visible light spectrum, attached mystical significance to the number seven, which is probably why he settled on that many markers. But the clue is in the name: the visible light *spectrum* is also a sliding scale, an endless stream of colours situated between red and violet that our eyes readily detect.

This chapter makes the case for the sliding scale – or, as mathematicians call it, the *continuum*. From a mathematical standpoint, a continuum is simply an unbroken line of numbers. That's *a lot* of numbers – infinitely many, of course. Yet even this description doesn't do it justice. Mathematicians have uncovered a surprising truth about numbers that reveals just how vast the number line is and, thus, how much more expansive the continuum is as a mental model compared to its discrete proxies.

'God made the natural numbers, all else is the work of man'*

Back to the doctor who, sensing my discomfort at matching my pain level to an emoji, followed up with a wink and a nod: 'Look, Junaid,' he said, 'I know you're a numbers guy, so let's ditch the faces and think of it as a scale from 1 to 10. And you don't even need to give me a whole number. Use a fraction if you prefer!'

I appreciated the gesture, which in practice gave me ample options. Yet the purist in me wanted to tell the good doctor that even allowing for fractions, my options

---

* This quote is attributed to the nineteenth-century mathematician Leopold Kronecker.

remained paltry. Fractions are not all there is on the number line – in fact, they barely feature at all compared to other number types.

Think about all the numbers you know, and the order in which you encountered them. First came the whole numbers – tangible quantities like 1, 2, 3 – what mathematicians call 'natural numbers'. Your number line soon extended to accommodate zero, and also the negative numbers (−1, −2, −3, etc.) that head in the opposite direction.

As you learned to divide numbers, you realised that a new type of number was needed, because, for instance, dividing 6 by 9 does not leave you with a whole number. Between 0 and 1 alone we can slot in numerous fractions: we can slice that segment into halves, or thirds, or quarters. We could, if we so desired, slice it into seventy-sixths, marking ticks at each new point. The fifth tick along represents the fraction $\frac{5}{76}$, the seventeenth tick marks $\frac{17}{76}$. And so the number line, which accommodates every one of these conceivable fractions, suddenly feels very concentrated indeed.

It's tempting to think that's our lot – that the number line comprises *only* the fractions. This was the conventional view among such luminaries as the Greek mathematician Pythagoras, for whom all quantities in the universe could be thought of in terms of a proportion. In this view of the mathematical world (a view apparently adopted by my doctor), all numbers are fractions.

But it turns out there are some numbers that cannot be expressed as fractions, however hard we try. The best known is everyone's favourite constant, pi (denoted by the Greek symbol π), the ratio of any circle's circumference to its diameter. This may sound like a fraction, but there are no two *whole numbers* whose ratio is exactly π. One of the main characteristics of these so-called 'irrational numbers' is that their decimal expansion will never repeat or terminate. The usual approximations of π, such as 3.14, are easy enough to pin down because they are expressible as fractions (in this case, just slice the interval between 3 and 4 into 100 equally sized smaller intervals and find the fourteenth tick along). But to capture a number like π in all its glory, it would not suffice to carve your number line into whole numbers, or hundredths, or even thousandths or millionths. You would have to carve up the number line repeatedly, infinitely. This hints at the incredible scope of the continuum.

So every number is one of two types: a fraction or an irrational number. Between any two numbers, you can find an infinite number of fractions and an infinite number of irrational numbers. Perhaps our comparisons should end there, but to truly grasp the extent of the continuum we must go further. One of these infinities, as we'll see, is larger than the other.

Our intuitions might lean towards thinking there are more fractions than irrationals because, as proportions, they seem more tangible than oddities such as π. And for practical real-world problems, we never need all those infinite digits anyway – NASA engineers do just fine with fifteen digits of π. The Pythagoreans could not countenance that such numbers could even exist; in their defence, the irrational numbers do not exactly jump out at us. The fractions must surely be the dominant force on the number line. Yet the exact opposite turns out to be true. Fractions might seem incredibly populous, but they can be placed in a single list. There's no way to order them in size, but they can be listed in a way that leaves no one out – they are said to be *countable* (an example of where mathematicians seem to have dropped the ball on terminology; *listable* would make more sense).

The method for listing the fractions is as simple as it is clever: we can first generate a grid of all possible fractions by making the rows correspond to the numerator (the fraction's top number) and the columns the denominator (the fraction's bottom number). The fraction in the sixth row and eleventh column, for example, is $\frac{6}{11}$. We have all possible fractions (some of them multiple times, in fact) in our grid. If we move diagonally through the grid, starting in the top left corner, we tick off every item in the grid – not a single fraction is missed.*

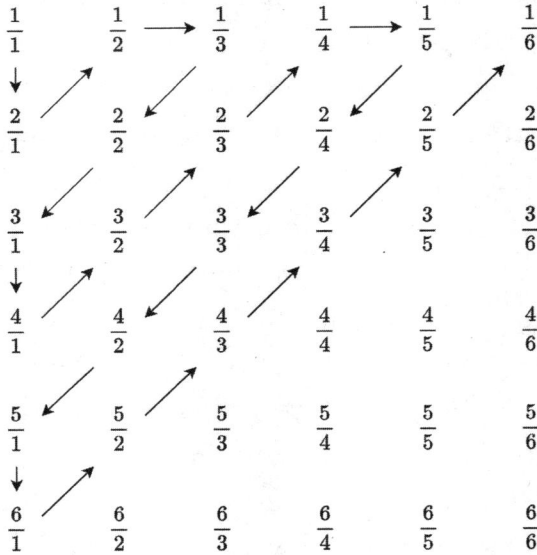

$$
\begin{array}{cccccc}
\frac{1}{1} & \frac{1}{2} \rightarrow \frac{1}{3} & \frac{1}{4} \rightarrow \frac{1}{5} & \frac{1}{6} \\
\frac{2}{1} & \frac{2}{2} & \frac{2}{3} & \frac{2}{4} & \frac{2}{5} & \frac{2}{6} \\
\frac{3}{1} & \frac{3}{2} & \frac{3}{3} & \frac{3}{4} & \frac{3}{5} & \frac{3}{6} \\
\frac{4}{1} & \frac{4}{2} & \frac{4}{3} & \frac{4}{4} & \frac{4}{5} & \frac{4}{6} \\
\frac{5}{1} & \frac{5}{2} & \frac{5}{3} & \frac{5}{4} & \frac{5}{5} & \frac{5}{6} \\
\frac{6}{1} & \frac{6}{2} & \frac{6}{3} & \frac{6}{4} & \frac{6}{5} & \frac{6}{6}
\end{array}
$$

The beginning of the list of fractions (excluding duplicate values) is:

$$
\frac{1}{1}, \frac{2}{1}, \frac{1}{2}, \frac{1}{3}, \frac{3}{1}, \frac{4}{1}, \frac{3}{2}, \frac{2}{3}, \frac{1}{4} \ldots
$$

Can we do the same with all *numbers*? Suppose I give you a list that I claim contains all the numbers. It will be

---

* Strictly speaking, this method only covers the case of positive fractions. But once they are shown to be countable, it is straightforward enough to modify the list to also include every negative fraction (simply place each one after its positive counterpart).

infinitely long, of course. Perhaps the first few numbers would look something like this, where each number is written as a decimal.

7.628423 …
0.500000 …
3.141592 …

The … denotes the fact that these expansions carry on forever. In the case of the second item, it is a never-ending sequence of 0s (we could just leave out the 0s, as they don't affect the value of our number, but you'll see why I'm including them shortly). The digits of the third item, which happens to be the number π, never repeat in this way.

If my list is the genuine article, I'll have shown that the whole number line is countable, just like the fractions. But if my list is not comprehensive, we should be able to find a number that doesn't appear on it. And it turns out we can indeed construct such a number, by dodging every item in my list as follows:

- The first digit is anything other than the first digit of the first number in my list – that is, anything other than 7.
- The second digit is anything other than the second digit of the second number in my list – that is, anything but 5.

- The third digit is anything other than the third digit of the third number in my list – that is, anything but 4.
- And so on.

As this is a number whose decimal representation disagrees with every number on my list (it will disagree with the eighteenth number on my list in the eighteenth digit and with the hundredth number on my list in the hundredth digit, for instance), it cannot therefore be on my list. Any attempt to construct an all-encompassing list of numbers will fail in much the same manner.

So the collection of all numbers is not countable. The fractions are not to blame – they are countable, remember. The problem lies with the irrational numbers; as unnatural as they may seem, there are simply too many of them to enumerate in a list. The extent to which they outnumber the fractions cannot be overstated – if you throw a dart randomly at the number line, you are not just *likely* to hit an irrational number; it is a mathematical certainty.

Let's take stock: in sizing up numbers, we have landed on four distinct sizes. Between 0 and 1 alone, there are:

- Two whole numbers – 0 and 1.
- Regularly spaced points such as tenths, hundredths or thousandths. Whatever our

choice of spacing, the number of points remains finite.

- All fractions – an infinite collection that can be enumerated in a list.
- All numbers (fractions and irrationals combined) – still infinitely many, but now they cannot be placed in a single list.

In quantifying just how 'many' numbers the number line plays host to, we've unravelled a new level of infinity – the collection of these numbers is 'uncountable'. There's nothing special about the segment from 0 to 1; any segment of the number line, however large or small, is a continuum. The whole number line (stretching forever in both positive and negative directions) is a continuum, but so is a segment between 0 and one-trillionth. That's largely what makes the continuum so vast; however much you zoom in, however tiny the segment, you still have uncountably many numbers at hand.

These four levels bring increasing degrees of resolution to our thinking. Let's take the previously mentioned example of consciousness. The first level represents the binary views of it as an all-or-nothing phenomenon where we apparently switch between conscious and unconscious states. But consciousness takes many forms such as daydreaming, drowsiness, hypnosis and sleep, each reflecting a different degree to which we are awake and alert to our surroundings. It's unlikely that consciousness can be boiled down to any finite number of states (the second level) or even a countable list of states (the third level). Since we appear to move fluidly between different levels of alertness, there is a strong case for thinking of consciousness as being situated on a continuum (the fourth level).[1]

Another distinguishing feature of the continuum that deserves brief mention is that it contains no gaps. The fractions do have gaps: between any two fractions there are many numbers that are not themselves a fraction. But between any two numbers on a continuum, every intervening number remains on the continuum – there are no sudden breaks.

This conforms to our real-world intuition. According to classical physics, motion is continuous: as I'm running, my centre of mass is moving through a continuum of points in space. Time is continuous too: even though we break the year into 365 days, and each day into 24 hours, time progresses through a continuum of moments.

This might seem an obvious point, but it eluded many of history's smartest minds. The Greek philosopher Zeno had no concept of a continuum, which led him to the paradox that bears his name. There are several variations of the paradox; the most straightforward observes that to travel, say, one mile, you must first travel half a mile, then half of what remains, and again, and on and on. Since there is always some distance remaining, Zeno's logic dictates that you will never reach the end – time and motion are but an illusion.

The continuum refutes Zeno's logic by permitting us to break the journey into an infinite number of steps, each one halving in size: half a mile, then a quarter of a mile, then an eighth of a mile and so on. So the total journey is given by summing these terms:

$$\tfrac{1}{2} + \tfrac{1}{4} + \tfrac{1}{8} + \tfrac{1}{16} + \ldots = 1$$

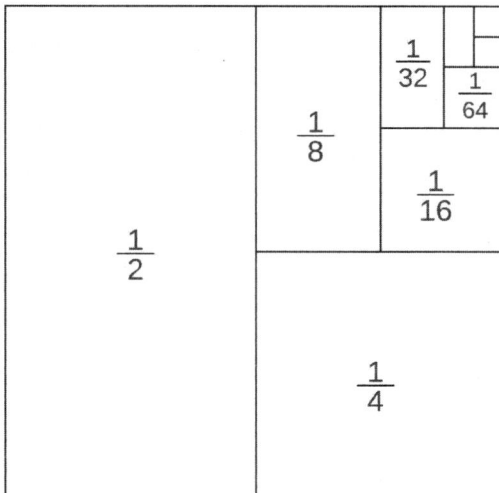

The infinite sum visualised with segments of a 1 × 1 square. Each fraction corresponds to the area of the smaller segment it is in. The sum of all (infinite) of these fractions is the area of the 1 × 1 square, which is just 1.

Never has an ellipsis carried such significance: this one means we keep summing the terms, indefinitely. It is an act of imagination that resolves our conundrum because we can see that this infinite process results in a total of one mile. There's an infinite number of terms to sum, but they are shrinking sufficiently quickly for the sum to *converge* on a value. The size of the steps halves each time, approaching zero while never quite reaching it. The steps can be thought of as an ever-diminishing line that increasingly resemble a single point, without ever quite becoming one. Zeno's blind spot arose only because he lacked the tools to apply a simple process – that of adding ever-smaller distances – over and over again, infinitely many times.

The core of calculus

The continuum takes centre stage in one of the most powerful areas in mathematics: *calculus*. At its core, calculus is the study of change – such as how an object transforms from one state to another, or how the trajectory of a ball changes as it moves through the air. The change may be imperceptible, but the continuum can cope with this because it can be broken into as many parts as we require.

It's a shame Zeno never got to meet Usain Bolt because the sprinter, perhaps more than anyone in recent human history, puts to bed the notion that motion is illusory. We can put a number to Bolt's exploits using the formula we learned at school: speed $= \frac{\text{distance}}{\text{time}}$. In his world-record-breaking run, Bolt covered 100 metres in 9.58 seconds, a speed of 10.44 metres per second (or 23.49 miles per hour). This estimate is crude, however, because it assumes a constant speed throughout. If we want to measure Bolt's speed in his absolute pomp, it would be better to ignore the early and latter stages of the run, where he is not at full speed. We might instead examine his speed at the halfway point, 50 metres in. If we consider the 40-to-60 metre range, we have a distance of 20 metres and it takes Bolt 1.67 seconds to cover that interval, for a speed of $\frac{20}{1.67} = 11.98$ metres per second (or 26.95 miles per hour).

Our measure of Bolt's speed at the 50-metre mark becomes sharper as we shrink the interval around it. Perhaps 5 metres either side of the 50-metre mark. Or just 1 metre. Or half a metre. Or less than that. And so on, by which I mean indefinitely. As we go through each iteration, the interval narrows towards a single point at 50 metres, without ever quite reaching it. Time approaches zero, but never quite reaches it. And Bolt's actual speed at the 50-metre mark is whatever the estimates converge towards, their so-called *limit*.

That, in a nutshell, is the basic idea of calculus: to break measurement into *infinitesimally small* parts.

We have effectively performed the calculation $\frac{0}{0}$, usually forbidden in maths, by approximating both the top and the bottom by values that approach zero, without ever quite getting there.

This method is routinely called on by mathematicians to measure the area of all manner of irregular shapes for which we don't have neat formulae. Suppose we want to estimate the area of this awkward-looking shape:

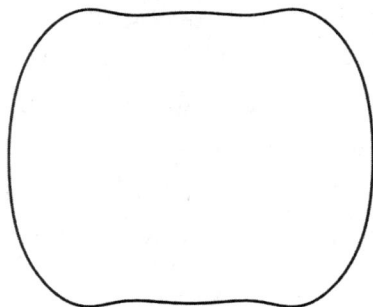

One way is to fill it with rectangles. We can easily compute the area of each rectangle and then find the sum. This won't give us the precise answer, of course; the rectangles stray over the shape's boundary, meaning we'll end up with an overestimate. We can do better by reducing the width of the rectangles. The process will be the same, but we'll have more rectangles and a better approximation of our shape's area. And, as you might have come to expect, we can continue this process forever, with the rectangles converging towards lines that collectively fill the space inside our shape and nothing else. Once again,

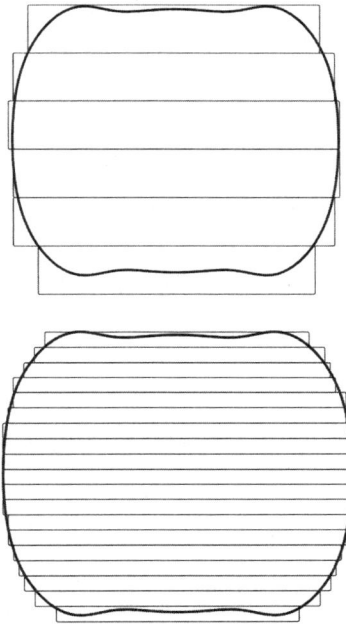

Two estimates of the area of our irregular shape. As the rectangles become thinner, the estimate becomes more accurate.

we never actually reach this point – the rectangles never actually become lines – but we can allow the process to go on and see where it leads (in this case, towards a precise answer for the area of our shape).

Invoking the continuum reflects a pointed approach to problem-solving that the mathematician and writer Steven Strogatz likens to an extreme form of divide-and-conquer. 'All good problem-solvers know that hard problems become easier when they're split into chunks', says Strogratz.[2] 'The truly radical and distinctive move of calculus is that it takes this divide-and-conquer strategy to its utmost extreme – all the way out to infinity.'

Back in the real world ...

Taken literally, the continuum does not apply to much of the real world. Who, after all, has the measuring tools (let alone the patience) to examine Usain Bolt's speed at the tiniest of intervals? Given practical constraints, such analysis would usually take place in 10-metre intervals, or 1-metre intervals for those seeking an extra degree of precision – a far cry from the infinitesimally small chunks of analysis offered by the continuum.

Many things we consider to be continuous are, in fact, anything but. The temperature dial on my shower purports to work on a continuous scale – in theory, I can toggle to my preferred level, with whatever degree of precision feels appropriate. In practice, that sweet spot is fiendishly difficult to find because the dial seems to operate at just three levels: punishingly cold, torturously hot and a tiny tolerable segment in the middle. The promise of a continuously refined temperature setting gives way to the most restrictive of morning routines, as I find myself confined to those three options.

While I'm in the shower, I may at least take solace in the continuous flow of water. Perhaps mercifully, I do not feel each individual droplet; instead, I am bathed in what feels like a smooth, uninterrupted flow. For the longest time, scientific consensus adhered to the view, popularised by Aristotle and Plato, that matter, like time and space, is continuous. It is only in the past couple of centuries, thanks to the scientific tools that have given us the means of probing matter at the atomic level and beyond, that the naivety of this view has been exposed. At the scale of individual water molecules, or atoms or quarks, there must be a point of indivisibility; our interaction with the world is fundamentally discrete.

Our perceptions of reality may not be completely accurate, but they are useful; if we processed every object in terms of its literal parts we'd never get anything done. Thinking of the world as continuous is a useful shorthand for the brain that means sacrificing accuracy for the sake of pragmatism, and the same is true for most mathematical models that invoke the continuum. When we speak of a population's growth, the long-term impact of economic policies or the spread of a highly contagious virus – phenomena that evolve in a discrete fashion – it helps to treat them as if they are continuous. By accepting these minor distortions, we create space to bring in the tools of calculus.

The main idea of the continuum is that zooming in to ever more minute scales reveals the truer nature of things. Physicists know this already; they have discovered that when we probe matter to ever smaller degrees a new world opens up to us. Exploring the world at a submicroscopic scale requires more energy, and things gets weird as quantum mechanical effects kick in. While we may be naturally inclined towards 'big' models of the world, new insights abound at imperceptibly small scales. This is not just true for physical phenomena. The continuum is also a portal through which we can make sense of human behaviours. It represents a philosophy of analysing events at infinitesimally small scales rather than accepting explanations at face value.

The calculus of history ... and our lives

Of all the references in this book, *War and Peace* might not be one you would have expected. Tolstoy's epic makes fleeting reference to calculus and demonstrates the power of the continuum as a model for analysing events.[3]

War and Peace is concerned largely with Napoleon's invasion of Russia in 1812. Tolstoy lays out a blistering critique, suggesting that Napoleon's influence on nineteenth-century European history has been exaggerated; his general point is that we ascribe too much influence to individuals – even leaders who wield great power and authority – on the events that surround them. His view is that history is too vast, too complex, to be explained by the actions of individuals.

Tolstoy makes his case through the lens of calculus – firmly established by this point, thanks to Isaac Newton and Gottfried Leibniz. He cites – and resolves – Zeno's paradox in much the same way that we did earlier; 'The movement of humanity', he writes, 'rising as it does from innumerable arbitrary human wills, is continuous.'

Tolstoy rejects any attempt to reduce history to a collection of discrete causes and effects, likening this paradigm to the geocentric view that places a stationary Earth at the centre of the universe. The historical endeavour, rather than focusing on 'arbitrary and disconnected' units of analysis, should take its cue from calculus:

> Only by taking infinitesimally small units for observation (the differential of history, that is, the individual tendencies of men) and attaining to the art of integrating them (that is, finding the sum of these infinitesimals) can we hope to arrive at the laws of history.[4]

This is 'integration' in the mathematical sense. Just as we computed the area of that awkward shape with ever more refined rectangles, using a continuum of measurements, history must similarly be indefinitely broken down into smaller slices for us to uncover its underlying laws.

At no point does Tolstoy offer a mathematical formula for 'computing' history, but that's the point – his aim is to move away from crass calculating tools, especially those that reduce the continuum of history to the machinations of a few people.

Your life need not play out like a Tolstoy epic for his ideas to register. The notion that our lives unfold one discrete episode at a time is appealing but misguided. The infatuated undergraduate we met in this book's introduction, who used an inductive model to predict an endless romance, fell afoul of this kind of thinking. Our relationships develop along a continuum of instances that bleed into one another. Love itself is a continuous construct; it waxes and wanes in infinitesimally small increments. As romantic as it may feel to pinpoint the precise moment that we fall in love with someone, the truth is that moments do not exist in isolation; they are entwined with all others that precede them. Our heart is not a switch; it may flicker but it never flicks from one state to another in an instant. Our emotions are situated on a continuum, steadily progressing in one direction or another. Just as we do not recall the moment we grew tall, we have to accept that love evolves in a manner that is not always perceptible to us.

If this is true of budding romances, it applies just as well to falling out of love. I'm fortunate not to have experienced this in my marriage, but it has occurred more than once in my career as roles that once brought fulfilment became unbearably tedious. It is tempting to attribute job dissatisfaction to singular causes – an overbearing boss, a failed project, a promotion that never was, an uncomfortable office environment – but the lesson of the continuum is that all these factors bear on our current state of being, as do countless others that we might overlook.

It is futile to search for a singular cause for our life-defining moments; our beliefs and actions reside in a melting pot of uncountably many experiences. To make sense of our life, we must zoom in to its smallest details.

The Intermediate Value Theorem

Look at the graphs below of two functions. You may be struck by one key difference. The graph below glides along very nicely, without any sudden jumps, much like the continuum – such functions are called *continuous*.

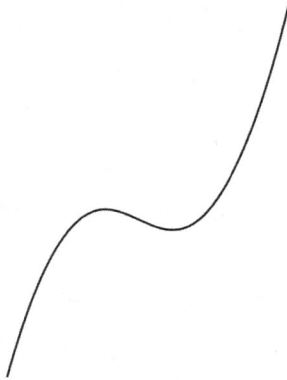

Their graphs can be straight, pointy or curved; the precise definition is rather technical, but the main idea is that the function never abruptly changes value. Now look at the following graph, which does experience a very sudden and obvious break – a so-called point of *discontinuity*.

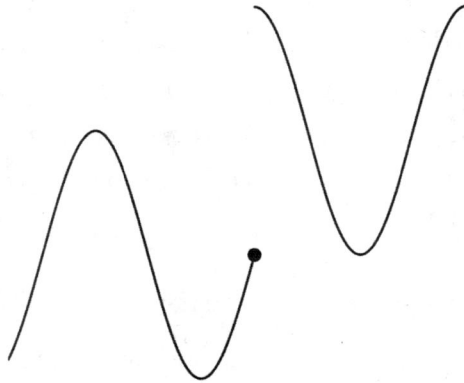

Mathematicians have much to say about both types of functions. We'll encounter a specific case of discontinuity in the next section. For now, we'll focus on a mathematical result for continuous functions – one that, among other things, helps me find my literal sweet spot when consuming hot beverages.

When offered tea or coffee at a friend's house, I will play the role of good guest and graciously accept. When asked how much sugar I want, I will assume the options are limited to a whole-number amount of teaspoons. If I overhear others trading in more refined amounts I'll follow suit, but I've never witnessed anyone asking for anything more precise than 'half a teaspoon' and I do not have the courage to break this etiquette.

In the comfort of my own kitchen, I feel no such inhibitions. You can imagine my wife's joy when I ask her to measure increasingly small amounts to help me pinpoint the exact portion. A mathematical result known as the Intermediate Value Theorem (IVT) gives us some assurance, at least, that this 'sweet spot' exists.

The theorem says that if a continuous function takes on a negative value at one point and a positive value at another, then somewhere between them is a third point where the function takes a value of zero. We can generalise by saying that if a continuous function takes on a value x at one point and a larger value y at another, then between those two points, the function takes on every 'intermediate' value between x and y.

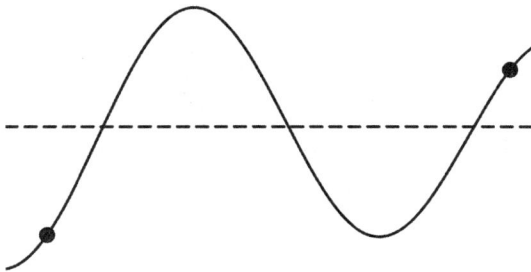

Here we see two points marked on a graph, where the graph takes on a different value (in this case a lower value for the left-hand point). If we choose any third value between these two values, we will always be able to draw a dotted line like the one above and find at least one other point where the graph takes on this third value.

One proof of the theorem offers us a step-by-step approach to locating our desired intermediate value. It is the same method we use at home to pinpoint my ideal portion of sugar. Zero sugars is too few and two sugars is too many, so we try the middle value – a single spoonful of sugar. I then make a judgement: too sour or two sweet? If it is too sour, my ideal amount lies between one and two sugars, so I try the next midpoint, 1.5 sugars. Likewise, if it is too sweet, then I try 0.5 sugars, the midpoint between 0 and 1. Suppose the single spoonful was too sweet but the 0.5 teaspoon was still too sour. I would then try 0.75 spoonfuls. If that was still too sour, I would head up to 0.875, and if that was too sweet I would head back down to 0.8125 spoonfuls. With each step my measurement is becoming more precise, and it is edging towards my ideal amount.

I can repeat this step ad infinitum (the power of the continuum once again) until I settle on the exact amount – and this is where the reality departs from the mathematics. While I can carry out the process infinitely many times in the mathematical realm, certain that it will culminate in the desired value, in the real world I would stop within a few iterations. For one thing, I am constrained by measuring instruments that cannot accommodate infinitesimally small gradations. But my willingness to search for the optimum amount is inspired by the knowledge that it lies somewhere in wait.

The IVT is one of those blindingly obvious results that mathematicians go to great pains to prove. But bear with me – some of the implications of the theorem are actually rather novel. We can use it, for instance, to show that at any moment there are two opposite points on the equator with the same temperature. To see why, consider any two diametrically opposed equatorial points. Suppose one of them, A, is warmer than the opposing point B. And suppose you took a trip to A, dispatching a friend to B at the same time. When you both arrive, you are in the warmer climate. The pair of you agree to move positions: you move clockwise from A to B, while your friend moves from B to A (also clockwise) at precisely the same speed. The journey ends when you arrive at B and your friend arrives at A. At this moment, your friend is in warmer climes; you have gone from being warmer than your friend to being cooler than them, and the difference in your respective temperatures progressed steadily along your journey (the continuity assumption). The IVT tells us that there must have been a point in the journey where you were no warmer or colder than your friend – the pair of points we are after.

Another application of IVT takes me back to my teenage years. My sister lived within walking distance of our parents' house, and had a pool table in her garage. It was only natural that I would pay her regular visits, usually under the guise of visiting my baby nephew.

Many evenings and weekends were spent smashing balls around the table, although a significant proportion of the time was devoted to vain attempts at trying to balance the table on the uneven garage floor. In the end I resorted to sticking folded pieces of card under one of the legs – an imperfect art that still caused the balls to veer towards one side of the table. It was the best I could do in the circumstances, or so I thought.

If I'd known about the IVT back then, I may have persisted in my search for a stable position. This application of the IVT requires an ingenious thought experiment.[5] It starts by assuming the four legs are of equal length. Regardless of how uneven the floor is, it is always possible for the table to rest on three of its legs, with the remaining leg hanging in the air. Call this leg A, and the others (in a clockwise direction) B, C and D.

Suppose we rotate the table 90 degrees, so that A is now touching the floor along with B and C, and it is D – in A's original position – that hangs in the air. Imagine the garage floor is made of sand, and push down on legs D and A, while keeping the opposite legs B and C fixed. Stop when D touches the floor, at which point A has extended below the sand.

We might not seem to be getting anywhere, but notice that we started with A dangling in the air and ended with it below the floor. And at both of these points, B, C and D were all touching the floor. This suggests there is a point somewhere that we can get A to touch the floor along with B, C and D, banishing the wobble in the process.

There is mathematical work to be done in justifying the use of the IVT – in particular, in modelling these moves in terms of a continuous function. That work was carried out thirty years after the problem was first posed.[6] Once again, the benefit of the IVT is not that it serves up a precise solution, but that it confirms the existence of one.

The usefulness of the IVT is primarily as an existence theorem – its purpose is to vindicate the pursuit of a sweet spot between any two positions we consider unsavoury. The IVT nudges us away from binary thinking, towards the continuum of intermediate possibilities. When I sample coffee that is too sweet, it would be folly for me to conclude that I'm just not a 'sugar person' when the issue is a matter of degrees. If I do not take well to a spicy curry, I would be severely restricting my culinary experiences by rejecting spices altogether. If my jacket makes me feel chilly but my overcoat too stuffy, I should not give up on clothing altogether – some layering will get me into my comfort zone. If a ten-kilometre run feels too toilsome for my body but staying at home leaves me restless, a shorter route may be called for. The IVT is a metaphor for balance; a reminder that between any two extremes countless possibilities abound.

Step functions

My driving route to the local swimming club is punctuated with traffic-calming measures. As I approach a particular set of traffic lights, I'm greeted with a digital face that either smiles at me to acknowledge I'm driving at a safe speed, or frowns to warn me I risk a speeding violation. These Smiley Activated Message (SAM) cameras exhibit a simple, unforgiving logic: any speed under 20 mph is deemed safe but as soon as I creep above it the judgement switches. This hardly seems fair, and only serves to make me an ultra-cautious driver, staying well below 20 mph to avoid the chagrin of the camera, which surely does nothing for road safety.

The SAM cameras exhibit a textbook case of *discontinuity*. The previous section, and much of this chapter, has concerned itself with functions without sudden breaks. As is often the case in maths, it is worth looking at the opposing concept. In this case, we have an example of a *step function*, where the value of our function (the SAM camera's facial gesture) is constant for a period and then suddenly rises or falls to another value.

Step functions are, at times, unavoidable. New parents learn this the hard way in the context of their sleeping patterns. The amount you sleep will not decline gradually; it will suddenly plummet. If continuity represents stability and predictability, there's no knowing where the fickle step function will head. Thankfully, we can learn to recover our sleeping habits and get back to more familiar routines.

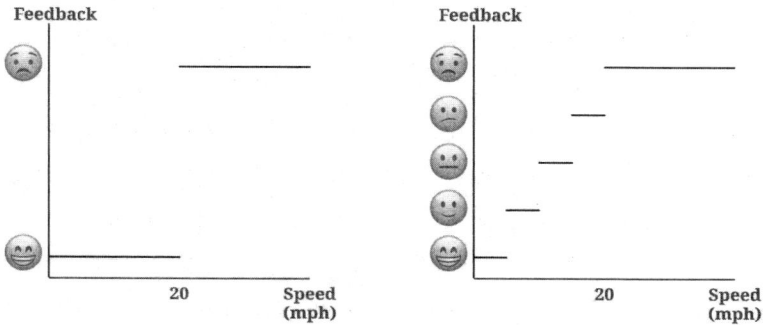

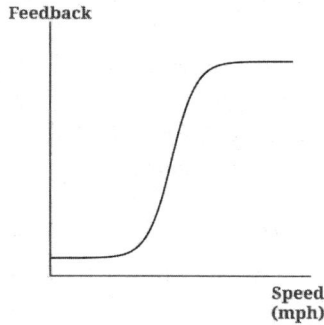

Left: a step function showing the two feedback states of the SAM cameras.

Right: adding more facial states reduces the vertical gap between any two successive states.

Bottom: a step function that has undergone smoothing and is continuous throughout, with no gaps. This is an idealised function that SAM cameras will increasingly resemble as more states are added.

But step functions occasionally reflect an inappropriate design choice. There is no reason why the SAM cameras should be confined to two states; no reason at all why they couldn't, say, be based on a steady progression of facial gestures. Happiness and satisfaction are not binary, and we can be trusted to interpret the *extent* of the system's approval based on the shape of its smile. Adding more facial states would have the effect of smoothing the relationship between a driver's speed and the feedback they receive. A small increase or decrease in speed will no longer be met with a drastic change in facial gesture.

A more consequential example of step functions comes from the way we define academic year groups. It is most unusual that two children born minutes apart could be placed a whole school year group apart, yet that is precisely what our conventions lead to. In England, a child born on 31 August – as late as 23:59 – will be placed in a year group with peers whose birthdays stretch back to 00:01 on 1 September of the previous year. A child born just two minutes later, at 00:01 on 1 September, will be elevated to the next year group. The first is the youngest in their peer group, the second the eldest in the next year group down.

Malcolm Gladwell's *Outliers* drew attention to the potential unfairness of this arbitrary cut-off point.[7] Looking at data from Canada (where the school cut-off is 1 January), Gladwell noted that a disproportionate number of professional hockey players were born between January and March and suggested that this had much to do with the physical advantage conferred on these students relative to their younger peers from the same cohort. The advantages may extend to academic outcomes too; a separate study shows that, in the UK, students born in August – the youngest in their cohort – are less likely to reach expected levels of learning compared with their peers.[8] As a September-born child, I acknowledge I may have dodged a bullet by getting a few months' head start in life over my peers.

Schools are constantly being challenged to adapt to children's individual learning needs, but very rarely does this discussion mention the variation that appears due to different dates of birth. Granted, there are no obvious solutions – it's not like the SAM camera, whose faults can be remedied by injecting more facial states. But we should be transparent with parents about the arbitrary nature of this cut-off point, and the potential for it to disadvantage younger students. Perhaps we should move away from rigid assumptions on when children should enter into formal schooling and give parents more options, without casting aspersions on late starters. And where we see step functions shaping our decisions, we should ask if the discontinuity – those sudden leaps – are justified, or if they just reflect lazy assumptions.

2

Gradients

Embracing life's ebb and flow

As each year draws to a close, my social media feeds greet me with the same productivity meme. You might have seen it yourself.

$$1.01^{365} = 37.8$$

$$0.99^{365} = 0.03$$

The calculations in these memes check out. They indicate plainly that if you multiply the number 1.01 by itself 365 times, your answer will be substantially larger than the answer you get by multiplying 0.99 by itself 365 times.

The deeper meaning behind this apparently mundane fact is signalled through the choice of 365 – we would, after all, arrive at the same conclusion with any positive value in its place. The calculation was popularised by James Clear's productivity bible *Atomic Habits* to highlight what can be gained – or lost – through consistent, small actions over a sustained period of time.[1] Taken at face value, it suggests that even a 1 per cent increase in one's effort, when applied day after day, results in significant growth over the course of a year. By contrast, a 1 per cent drop in effort amounts to a steep deterioration over the same period – mathematical proof that your New Year's resolutions are worth sticking with after all.

The meme is rooted in the idea of *compounding*, a term bandied about in investment circles, when interest is earned not only on the original payment but also on any interest that has already been paid. The same calculation is at play: a bank account containing £5,000 receiving 5 per cent annual interest increases its value by a factor of 1.05 over a single year. Over ten years, it increases by 1.05^{10} (that's 1.05 multiplied by itself ten times – one factor of 1.05 for each year), reaching a pretty sum of £8,144.

Compounding defies many of our intuitions. If you fold a sheet of 0.1-mm-thick paper in half, you obviously double its thickness. If you fold it again, the thickness will now be four times the original. Each subsequent fold contributes another doubling; a classic case of compounding. A popular question asks how many folds you'd need for the thickness to reach as far as the Moon (indulging the assumption that we can penetrate any physical limits around paper folding). Given that the distance is roughly 384,400 km, you could be forgiven for thinking we'd require hundreds, maybe thousands, of folds, yet it only takes 42. Another 52 folds would take the thickness to the edge of the observable universe.*

Einstein is alleged to have called compounding the eighth wonder of the world, as well as the strongest force in the universe. Evidence for both quotes (like so many that are attributed to him) is scant, but there is clearly something to the sentiment.

Mathematicians refer to such sequences, where terms are multiplied by the same factor each time, as *geometric*. The size of this factor determines what happens to the sequence. If it is greater than 1, the values become larger and the sequence eventually surpasses any given amount (in which case it is said to *diverge to infinity*). If the factor is between 0 and 1, the values get smaller, approaching but never quite reaching zero (they are said to *converge*). Our original example shows how sensitive this difference can be; a mere difference of 0.02 (0.01 either side of 1) results in markedly different outcomes.

Even though the productivity metric applies to 365 separate days, giving rise to 365 distinct values (from 1.01 to 1.01^{365}), the term 1.01^x is defined for *any* value of x (for our purposes, it is sufficient to consider non-negative

---

* The *practical* world record for paper folding belongs to a group of students who achieved thirteen folds with a 13,000-foot-long roll of toilet paper. To reach the Moon, the sheet would need to be billions of times as long, which surpasses the actual distance to the Moon.

values) and not just whole numbers. Likewise for 0.99^x. This means we can sketch one graph for 1.01^x, which helps to visualise its growth, and another for 0.99^x, which helps to visualise its decay. The contrast between the two is striking – we can see how the paths deviate from one another over time, depending on the value of the 'effort' factor.

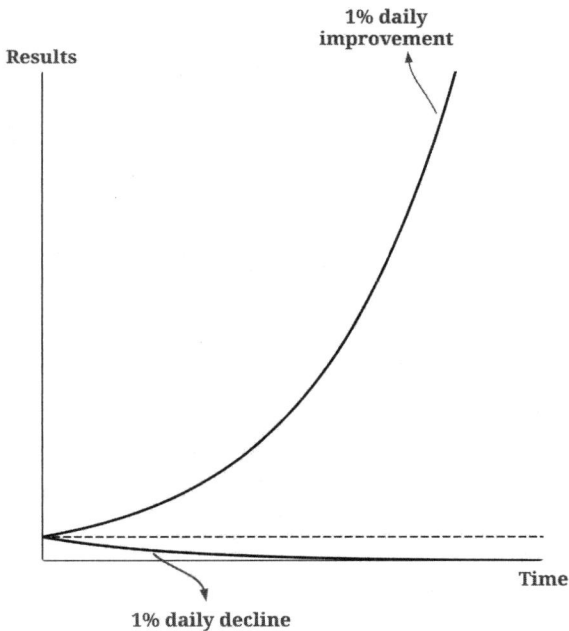

Productivity curves (adapted from James Clear's
Atomic Habits): when each day's effort is 1 per cent more
than that of the previous day, the curve shoots up to higher
and higher values. When each day's effort is 1 per cent less
than that of the previous day, the curve heads downwards,
approaching zero without ever quite reaching it.

Now that we have graphs in play – and smooth-looking ones at that – we can bring in ideas from calculus. In particular, we can use the concept of *gradients* (also called *derivatives*) to better understand the growth assumptions underpinning our earlier productivity calculations. For the purposes of this chapter, we will think of gradients in terms of a slope – how steep a graph is at any given point.

The simplest case is that of the horizontal line, which represents stagnation – the same level of effort each day. It is unfailingly flat with a gradient of zero at every point – no hint of a slope, no signs of progress in either direction. If we increase our effort, say by adding 1 per cent of our original effort level to each day's exertions, our line would now tilt upwards, with a constant gradient of 1.01.

This is not what the productivity meme is getting at, however. The 1 per cent daily increase it calls for is not additive but multiplicative. On each new day, we are not just adding 1 per cent of our original effort but, rather, 1 per cent of the previous day's effort, which itself was a 1 per cent increase on the previous day. In graphical terms, this is the difference between a straight line with constant positive gradient and a curve that shoots up dramatically.

You may recognise this as an example of *exponential* growth, the defining feature of which is that the gradient is proportional to the function itself. This means that as the value of the function increases, its gradient increases at the same rate. It is the continuous analogue of the wonder of compounding. It applies in the other direction too: a function exhibiting *exponential decay* has a negative gradient (its value decreasing as you move along the x axis), and as the value of the function decreases, its gradient decreases as well, at a proportional rate.

This concept gives rise to one of my favourite mathematical phrases. To express doubt in your assessment of a situation is a sign of humility, but there is no stronger way of conveying this than 'placing your uncertainty in the exponential'. To make sense of the phrase, consider the following two mathematical functions:

$$x^2 \qquad 2^x$$

Both contain an unknown, x. In the first expression we are multiplying x by itself. In the second expression, it is the number 2 that gets multiplied by itself – an x number of times.* A slight change to the value of x will inevitably change the value of each expression, but the effect is more pronounced in the latter, where the x is playing a weightier role. If x changes from 10 to 11, for instance,

---

* This assumes x is a whole number, although the expression can still be defined for non-integers using the logarithm function. The details are omitted here.

the first expression, x^2, changes from 100 to 121 (a factor of 1.21), whereas the second doubles (a factor of 2). In terms of gradients, the graph of 2^x exhibits classic exponential growth, whereas the graph of x^2, while rising at ever-faster rates, does not come close to matching the growth of its counterpart beyond the smallest values of x.

To place your uncertainty in the exponent is to acknowledge that if your assumptions regarding a particular situation are even slightly amiss, your conclusions may be way off the mark. Making any kind of prediction under such conditions carries reputational risk, and it's something pollsters reckon with every time they forecast election outcomes. An individual factor – such as how heavily the economy weighs on voters' minds – can have an outsized impact on the result. And because no poll can precisely capture voters' true opinions, pollsters are at the mercy of small miscalculations that can dramatically distort their overall prediction, and have no choice but to place their uncertainty in the exponential. The same goes for entrepreneurs who exhibit a strong appetite for risk when starting a business and are willing to entertain the widest range of outcomes based on factors they cannot claim to have control over. Their pitches may exude seemingly unshakable confidence, but their uncertainty too resides in the exponential.

Fluctuating fortunes

I can't help but wonder if the productivity gurus have missed a trick with exponential growth. Why not instead encourage us to double our effort each day? Within ten days, our output will have increased a thousand-fold. Within a month, by more than a billion. We'd need new nomenclature: *celestial habits*, perhaps.

The notion is preposterous, though no more so than the atomic variant that has recently gained prominence. It is based on the underlying idea that our effort and output is sustainable over prolonged periods of time, but the Covid pandemic brought home just how facile this assumption is.

As society began to lock down during the first waves of Covid-19, the gurus exhorted us to take advantage of our new, socially distant reality. To them, the solitude imposed by lockdown was merely an opportunity to devote more time and energy to our work. In place of a daily commute and in-person meetings, we could all – if we only put our minds to it – climb the productivity curve. Held up as the exemplar was none other than Isaac Newton. It was during his 'lockdown' from the plague in 1665, from his family estate of Woolsthorpe Manor in Lincolnshire, that he enjoyed his 'year of wonders', developing his theory of optics and (appropriately for this chapter) calculus.

I wonder how Newton might have adapted to the demands of modern society. How would he have dealt with back-to-back Zoom meetings, in the presence of children wondering why school was suddenly off limits? The measure of the pandemic, for many of us, was hardly predicated on continual growth. Any gains in productivity were readily offset by the hours of social media doom-scrolling, not to mention the constant distraction of prognosticating about the toll the virus might take on us and our loved ones.

If there was a silver lining to the pandemic for me as a mathematician, it is that it firmly planted the concept of exponential growth in the public psyche. With case numbers doubling at regular intervals (every ten days or so during the initial onset of the pandemic), the exponential function made its way into watercooler chats (at least until we were all working virtually), as pundits, politicians and citizens of every description scrambled to make sense of the surge in cases and debated the impact of public health measures.

Yet the exponential model of viral spread only tells the first part of the story because such trends are difficult to sustain in the real world. When the economist Thomas Malthus warned in 1798 that the human population would eventually outstrip our agricultural yield,[2] he did so on the basis that population rises exponentially, whereas agricultural growth nudges along at a steady clip. The result would be an overbred, underfed society languishing in poverty. But Malthus had not countenanced that the global population would level off. As society becomes more prosperous and people escape extreme poverty, they usually have fewer children because they are less dependent on child labour to supplement their income. Between 1948 and 2017, the average number of babies per woman halved, from five to just below 2.5.[3] This means that although the global population will continue to increase for a while, the rate of growth will slow as new births no longer overwhelm the number of deaths.

The exponential trajectory of the pandemic relented for similar reasons. Over time, as more people become infected there are fewer people left in the population for the virus to spread to, meaning that new cases must plateau at some point. Even as immunity wears off and new variants emerge, the combined effects of public health measures and breakthroughs in vaccines and drug treatments take hold. The story of the Covid pandemic is one of ebbs and flows, each new spike in cases being met with a raft of measures designed to flatten the curve and bring it back into descent.

This brings me to some lessons that we might have learned during the pandemic – but that went largely unnoticed by productivity gurus. The first is to do with the fluctuating curves of long-term pandemic forecasts, which approximate our productivity habits more closely than exponential curves do. When I took up running in my twenties, my progress was steady at first and then, over an exhilarating several-week period, the gains were sharp. During this time, as my distance and speed increased in equal measure, my potential seemed boundless. I was rising up an exponential curve, with no end to my progress in sight. Eventually – and inevitably – my performance gains diminished and any improvement in my times became incredibly hard fought as I reckoned with my physiological limits.

This pattern of progress is reasonably well described by an *S-curve*, so named because of its shape. An S-curve conveys three phases of growth: steady take-off, a period of rapid improvement, and steady growth as a limit is approached. As novices, assimilating to a new skill may feel laborious initially, but if we persist something will click and we can delight in rapid gains. The price to be paid for mastery of any craft is that top-end gains are so much harder to come by, which is why we tend to find ourselves stuck in the final portion of the curve, where the effort required to achieve even the smallest gains can seem daunting.

An S-curve approaches a limit without ever quite reaching it, a part of a curve that mathematicians refer to as an *asymptote*. In this context, it is a recognition that there is always further progress to be made. One never masters a skill completely – runners can become faster, musicians more expressive, entrepreneurs richer,

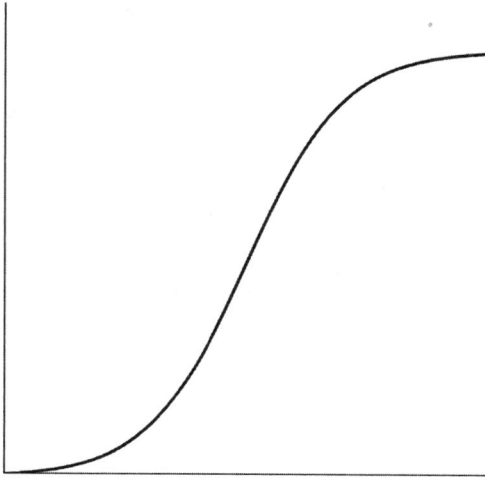

An S-curve: steady take-off, rapid gains, steady gains.

but perfection is an unattainable ideal. 'Have no fear of perfection', Salvador Dali advised. 'You'll never reach it.' Samuel Johnson wrote that it is 'reasonable to have perfection in our eye, that we may always advance toward it, though we know it can never be reached'. The S-curve is a humbling reminder that progress eventually stalls.

In any pursuit, it is helpful to have an idea of where you are situated on the S-curve. The gradient at any specific point indicates how much you stand to gain from a given quantum of effort. Running remains an important part of my life, but the level of effort I now put in is determined by where I think I am on the curve. If an injury or bout of illness prevents me from lacing up, my motivation to hit the roads once I've recovered is sky-high – I know I can look forward to rapid gains in fitness. During periods of peak fitness, however, I have to double down on things like diet and conditioning if I'm to improve, and even then the gains are limited.

S-curves crop up in many places, including in the adoption of new technologies. The first phase represents the 'early adopters', enthusiasts who are willing to look past the rough edges of a new product. As the technology improves and reaches more people, adoption rates rise until it achieves market saturation and sales plateau. Businesses that are subjected to the S-curve trajectory must stave off the threat of long-term stagnation and decline through innovation, diversifying their offerings to reach new markets – and thereby triggering new S-curves.

This may also explain the overarching progress of technology: as incumbent technologies grind to a halt, new innovations take their place – at least for a while, until they too are superseded. The result is a succession of overlapping S-curves that guarantees the upward mobility of technology.[4]

This model suggests that even as we plateau in one discipline, rather than resign ourselves to perpetual stagnation we might use our skills to fuel our development in other areas. Diversification and innovation apply to individuals too. Middle-distance running is only one proxy for exercise; I've barely scratched the surface when it comes to my all-round fitness. There is unexplored territory in the form of aerobic exercises such as cycling and swimming (which I undertake casually, and which are rooted at the steady front end of their respective S-curves), and core strength training (which I've barely invested any effort in). Perhaps there are also new disciplines, seemingly untethered to running, where the qualities I've developed as a runner put me in good stead. Our most valuable skills and habits are the ones that permeate the boundaries of different disciplines, allowing us to transfer success in one field to a seemingly unrelated one.

The S-curve still falls foul of assuming positive growth throughout; the gradient may vary, but the curve always remains on an upward trajectory. But this isn't how life unfolds. Even some of society's most revered minds, who would surely be considered productive by any reasonable standard, are susceptible to the occasional off day, or indeed whole periods, when progress stutters. Wittgenstein's diary contains a succession of entries beginning with the words 'Did no work', followed by an honest appraisal of his dulled state of mind. 'Will I ever work again?' asks one entry, while another declares: 'I have lost all hope and confidence in my power to succeed.' A diary entry from Franz Kafka, meanwhile, pointedly states that the novelist is experiencing a 'Complete standstill'. His malaise endures a good while; the following month he proclaims: 'How time flies; another ten days and I have achieved nothing.'[5]

What's true of writing is true of any creative pursuit – our best work is often preceded by periods of necessary struggle. It is certainly true of mathematics, where the most taxing problems are rarely solved in a smooth, upwards fashion; breakthroughs often only come about after one has reckoned with the toil of failed attempts and resigned oneself to defeat. I experienced this first-hand during my PhD when, after three years of stagnation and false starts, I enjoyed a six-week purple patch in which all my breakthroughs arrived in quick succession.

It is with some scepticism, therefore, that I engage with productivity tools that promise to remove the friction of creative work. Nowhere is this more pronounced than with the advent of AI-powered 'writing assistants' that purport to alleviate writer's block. As appealing as these may sound, confronting the blank page and enduring some degree of struggle is a prerequisite for expressing our most authentic and original ideas. To be productive, in the creative sense, means allowing for stasis.

There are times when we simply don't have a choice but to slow things down. The Covid pandemic served up the most unwanted reminder of our vulnerability to forces beyond our control. When all it takes is a debilitating virus to grind our daily routines to a halt, we should call out the culture that demands we plough on. Productivity models predicated on perpetual growth are hubristic at the best of times; during times of crisis they are untenable.

We need models that allow for stagnation and periods of decline alongside growth. Mathematics has much to offer in describing the erratic undercurrents of life.

Stationary points

Mathematics is replete with its own vocabulary, though thankfully much of it makes intuitive sense. A point on a curve where growth is neither positive nor negative is called a *stationary point*, and under certain conditions such points are inescapable.

We are about to encounter a mathematical result that captures a truth that is plainly obvious on a moment's reflection but that often gets lost in our attempts to analyse the vicissitudes of life.

Rolle's theorem* asks you to imagine that a curve takes the same value at two different points. We also assume (as we have been doing throughout this chapter) that the curve is smooth enough for its gradient to be measured.

---

* Incidentally, the original proof of Rolle's theorem took a more algebraic approach as in his early career, the mathematician, whom the theorem is named after, railed against the infinitesimal paradigm of calculus.

Then, according to the theorem, a stationary point can always be found between these two points.

Rolle's theorem may not be a stunning revelation, but it does keep us attuned to the inevitable highs and lows of our existence. After all, any journey that ends where it started must allow for at least one momentary pause as you wind through life's twisty paths.

Notice how the gradient changes sign as you pass through a stationary point – it either switches from

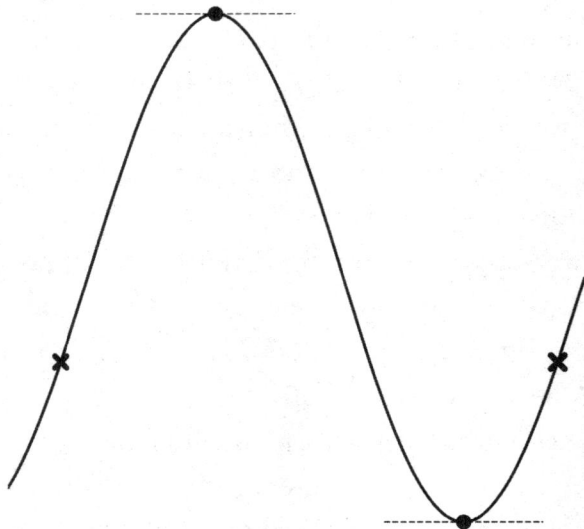

The curve has the same value (height) at the points marked **x**.
In this case there are two stationary points between them
(namely, the two marked dots) where the curve has a gradient
of 0. Rolle's theorem guarantees that at least one such
stationary point will *always* be found in the interval between
two points where the curve has two matching values.

positive to negative, or vice versa. The first case applies to situations in which you are punching above your weight, exceeding your usual performance levels until you hit a peak. Thereafter, a period of decline awaits you as you regress back to your former self. It is as Newton said: what comes up must come down. What he neglected to mention was that the converse holds as well. At other stationary points, the gradient may pass from a negative value to a positive.* If you find yourself on a downslope but have your sights set on recovering past glories, you may first need to accept that things must get worse before they get better. At some point you will sink into your lowest ebb. There you will remain, before climbing the pits of despair. What goes down must come up.

Rolle's theorem is a special case of a more general result known as the *mean value theorem*.† This time we do not require that our curve ends up where it started – it may end up at a higher value or a lower one. Consider the straight line that connects these two points. It will have a

---

* The exception – too mundane for our purposes – is completely flat lines, where the gradient is zero everywhere.

† Mathematical results sometimes get named after the people who first discovered them and, evidently, sometimes they don't. Sometimes the result is attributed to the wrong person altogether. The French mathematician Augustin Cauchy proved the mean value theorem, although he more than made up for it by stealing credit for other results that are unfairly named after him.

gradient of some description (positive if the second value is larger, negative if it is smaller, zero if they are the same, as is the case with Rolle's theorem). This gradient can be thought of as the average rate of growth in that interval. The curve may behave in various ways between the two points: it may rise or fall, gradually or wildly. At various points, the curve's gradient may be positive or negative, large or small, depending on its growth patterns.

The mean value theorem tells us that, irrespective of the shape of the curve, there is a third point somewhere in the interval where the gradient matches the long-run average rate of growth (the 'mean value'); in other words, it matches the slope of the line connecting the two points.

The theorem can be illustrated by the 5-kilometre Parkrun, a weekly ritual of mine. The organisers implore us to treat the event as a run rather than a race, yet they also insist on timing and ranking every participant. A runner's progress can be described with the classic distance–time graph, where the gradient at any moment corresponds to their speed.

My usual pattern is to come out of the blocks flying, lose steam as the run progresses, before finishing with a breathless sprint. Suppose I've set myself the target of completing the race in 22 minutes, equating to a respectable overall speed of roughly 13.6 km/hour. Assuming I have a steady run, the mean value theorem guarantees that somewhere along the journey, between my initial

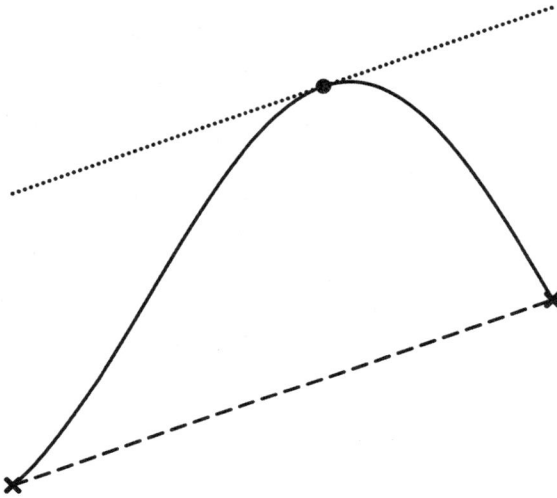

Here we have two points, marked **x**, where the curve has two different values (heights). The gradient of the dashed line shows the overall rate of growth of the curve between these two points. According to the mean value theorem, there is guaranteed to be a third point in the interval between these two points where the gradient of the curve matches the gradient of the dashed line. Here the third point is marked with the dot and we can see that the gradient of the curve at the dot, given by the steepness of the dotted line, is the same as the gradient of the dashed line (the two lines are parallel).

burst and final sprint, I will hit that speed exactly, however momentarily.

Relative to our long-term rate of progress, it is not only *acceptable* to have moments of 'average' performance; the mean value theorem tells us it is *inevitable*. When you look back on your progress towards a goal, it's easy to focus on periods of dramatic productivity or to dwell on the slumps. But there will also be moments between those extremes, when you were simply average. It is a mathematical certainty.

The mean value theorem may border on truism, but having this insight in advance of achieving our goals can be liberating. It may take the formality of mathematical analysis to remind us that not every moment in life needs to be – or indeed can be – experienced in terms of peak performance. But still, let's head to those peaks and ask what it means to optimise our fortunes.

Gradient descent and the local optimum

Picture yourself scaling the summit of a mountain with the weather conspiring against you. Thick fog surrounds you, impeding your visibility to the extent that you are unable to survey the landscape. Google Maps is out of the question (let's say you are in a rural region with poor phone reception). How do you find your way to the top?

This is a moment when your sense of touch will be your guide. You walk around in a small circle, gauging the steepness at each point. Identify the spot with the steepest upwards slope and head in that direction. Stop and repeat, each time taking a few steps upwards. There's a reasonable chance that this method will lead you to the summit. Google Maps be damned!

The process you have just followed – the algorithm – is known as *gradient ascent,* and it is remarkably powerful in areas such as AI, where machines seek patterns in large datasets. You may recall being given a scatterplot of points at school and being asked to draw a line of best fit. You probably just took a pencil and drew a swipe, satisfied that your attempt was close enough. But unless the points are situated on a line (which would render the exercise moot), you'd have to accept some degree of error. Wouldn't it be great if there was a procedure you could use to systematically minimise the error?

This idea is at the heart of machine learning. Of course, the models are more complex, typically involving billions of parameters and far more sophisticated curves than the line you drew. But much of the implementation of these models amounts to fitting curves into these high-dimensional data clouds. Whenever a machine learning program makes a mistake, its error score goes up, and we want the curve that minimises the overall amount of error.

As the goal this time is to minimise something, it's more akin to climbing down a valley while seeking the lowest point that leads to your escape. The algorithm we require is gradient *descent,* and the idea is much the same as before. Much like your mountaineering self, the algorithm will start at a particular point and then automatically take a step in the direction that most dramatically reduces the amount of overall error. It lands at a new point and then does the same thing over and over again, until it is no longer able to minimise the error – at which point we've found the parameters that correspond to the minimum amount of overall error in the curve.

Gradient descent (and ascent) is a homage to trial and error. Some of the most intelligent machine and human behaviour results from assessing our surroundings and opting for the move that nudges us towards our goal. If we do that repeatedly, we stand a good chance of reaching our desired endpoint.

A good chance, however, is not a cast-iron guarantee. If the fog clears, joy might turn to lament as you realise that there are other summits further ahead, reaching heights greater than yours. You are in what mathematicians call a 'local maximum'. Gradient ascent has led you to the highest point in your immediate vicinity, while leaving you oblivious to the loftier summits nearby.

We can easily find ourselves in the trappings of the local maximum. As a student, I struggled to keep up with conversations around career planning. My peers would use terms that were alien to me; they would speak enthusiastically about 'milk round' events, and internships at the 'big four' accountancy firms. It struck me that, for all the diverse talents surrounding me, a small handful of vocations dominated the aspirations of my cohort. Management consultancy and investment banking made up the lion's share of job prospects among my course mates, with a majority of career-related pamphlets beckoning me towards one of these paths. While I resisted (mainly because I was too occupied with maths problems and pool tournaments), a good majority of my fellow students landed themselves in one of these roles – initially as interns, and then as full-time employees on a fast track to highly paid and prestigious careers.

Yet when I caught up with my friends, they would regale me with sad accounts of how badly their jobs were progressing. Eighteen-hour days were not uncommon, often combined with a lack of passion – or even basic enjoyment – of their day-to-day work. As someone who places huge stock in a balanced lifestyle, I could not fathom how these talented and committed people could end up in such unfulfilling roles. Some would speak in defiant tones, convincing themselves they had opted for the right career path.

It was clear to me that my friends had been seduced by the pervasive narrative that the careers best suited to ambitious, intelligent twenty-somethings are those that demand the most time and energy, and that promise monetary and reputational rewards above all else. These tropes are often perpetuated by pressure from senior family members and wider society, who place a premium on traditional career paths that offer prestige and success early on.

For some of my peers, these careers were a genuine match for their skills and interests, but it was evident that others surveyed the narrowest spectrum of options and found the most palatable among them, remaining oblivious to more fulfilling opportunities beyond their search criteria. They were trapped in a local maximum. An Oxbridge degree is itself a local maximum, by the way: many students convince themselves they've already reached an intellectual summit and use it as justification for narrowing their career options or academic pursuits.

It is not just recent graduates who succumb to the local maximum; the phenomenon can crop up throughout one's career. As we settle into a role and develop an attachment to our employer, our view of what constitutes meaningful and rewarding work becomes increasingly insulated. It is perfectly natural to feel jittery after occupying a role for several years and wonder if there are more challenging and rewarding opportunities elsewhere. But leaving a role is another matter; for many people, the known quantity of an existing employer is preferable to the myriad unknowns that reside beyond company borders. I found myself in this situation five years into a role that had clearly reached its expiry date in terms of my professional development. Yet I hung on for another two years, switching roles and responsibilities in the hope that changes within the company would quell my concerns. Those two years were the least productive of my career. I was seeking to optimise my circumstances within an organisation that was no longer compatible with my career ambitions. I was trapped striving for a local maximum that, even if achieved, left me demanding more of myself.

The pressures of family and culture can also assert themselves, never more so than in our search for life partners. In too many communities, intense scepticism is reserved for partners whose families herald from a different country (or, in more extreme examples, from a different village within the same region of the same country!). As someone with Pakistani heritage who is married to a Libyan, I experienced these attitudes first-hand. While our story had a happy ending, I've seen many friends succumb to the notion that one's country of origin is a key marker of compatibility. They would even have you believe that degree choice is predictive of marital success – as if only medics 'get' other medics, or only a mathematician could handle the analytical disposition of another. It is no more enlightening than the view that your soulmate must reside in the village where your parents grew up. And as with careers, we can fool ourselves into believing we've made optimal choices having severely restricted our search parameters.

So what steps can we take to boost our chances of escaping a local maximum and reaching the actual summit – the global maximum? Giant ones. Rather than nudging yourself a few steps along in every increment of the gradient ascent (or descent) algorithm, you could trek much further in a random direction, leaving you in some other part of the terrain. When you stop, you can get back to diligently applying the algorithm that gets you to a local summit. If you repeat this exercise, you might hit several local peaks. The largest among them may not be the global maximum, but chances are you are at a higher peak than you'd reach with a single application of gradient ascent.

In the search for career fulfilment, this method may translate to applying for jobs outside your current organisation. It is a leap into the unknown, with no guarantee of success. But the wider your search criteria, the better your chances of landing a role that is more closely aligned with your values and career goals.

The same principle applies to relationships: your chances of finding the ideal partner increase manifold once you cut loose from societally imposed restrictions. A willingness to explore new frontiers, and to break free of imposed boundaries, can pay dividends. You may need a push along the way, but this is what good friends and colleagues are for.

And remember, the path towards a global maximum is undulating: when you are at a local maximum, any change of direction will initially take you downwards. You may have to fall before you rise and trudge your way through life's valleys before you eventually reach new heights.

Thinking in second (and third) derivatives

Returning to the trajectory of pandemics, which of the following two points feels more reassuring?

A cursory glance at what follows each point should settle any debate. In the first case we're approaching a plateau, a welcome contrast to the sharp uptick we are faced with in the second. In a different context, where the segments represent parts of a productivity curve, the second option feels more exciting, with rapid gains in store.

In terms of their gradient, there is nothing to tell the two points apart. Our reasoning just now took things a step further by examining *how the rate of change was changing*: 'plateau' is shorthand for a gradient in decline, while those rapid gains can be understood in terms of an increase in the gradient.

We are now in the territory of the gradient of the gradient, also known as the *second derivative*. Consider the following distance–time curve, which is growing quickly.* If we take the gradient at each point and plot these values on a new graph, we end up with a speed–time graph. And if we do the same thing again – taking the gradient at each point of the speed–time graph – we end up with an acceleration–time graph. In this example, acceleration is the second derivative of distance with respect to time (and it is flat in this example because the speed–time graph is linear, implying a constant gradient).

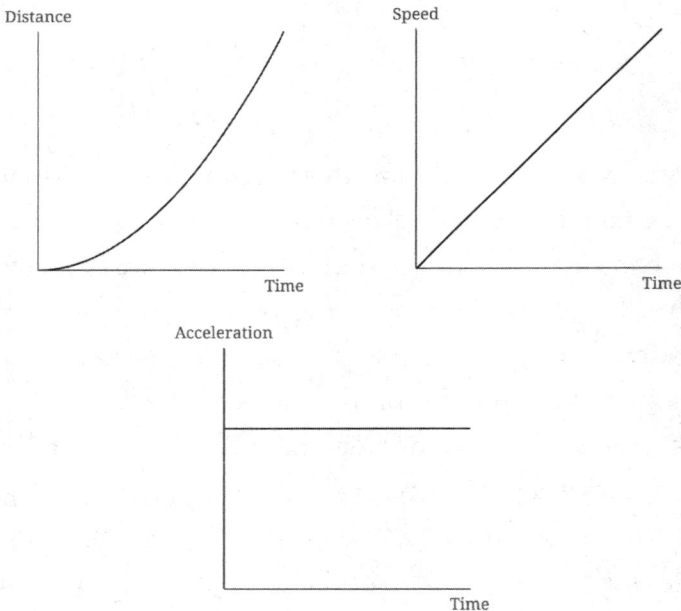

_____

* Quadratically, to be more precise.

The second derivative is a measure of a graph's curvature. In simple terms, a positive second derivative means that portion of the graph is curved in such a way that the opening is facing upwards. The gradient is on the rise. When the second derivative is negative, the opening faces downwards and the gradient is on the wane.

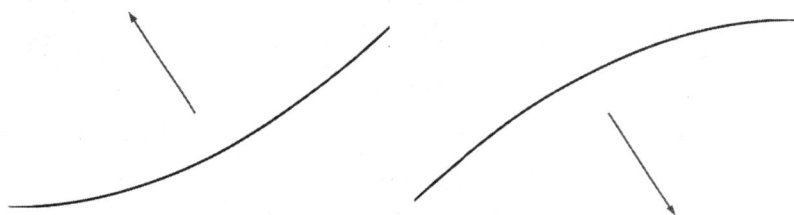

There are certain *inflection points* where the curvature changes from one of these states to the other. If we know the shape of a curve, we can identify these points by seeing where the second derivative transitions from positive to negative (or vice versa). On an S-curve, for instance, the inflection appears in the middle portion, somewhere on the steep incline.

Knowing that you are at or close to an inflection point can pay dividends. In the adoption of new technologies, it is the point where a new product reaches the mainstream, and where the growth rate in user uptake peaks. Your urge to invest in these technologies may be tempered by the recognition that growth is about to slow as the adoption curve heads towards its plateau.

Far better to invest in the early stages, where growth is not impressive but the rapid gains are still to come. In a pandemic, the inflection point can be viewed in optimistic terms because it suggests that the rate of spread is about to slow, which may inform a relaxation of public health measures.

Second derivatives also feature in political discourse. Politicians who promise to slash the rate of inflation are invoking second derivatives because inflation already represents a rate of change – the rate at which the price of goods has increased. Contrary to the common misconception, far from reducing the price of goods, a reduction in inflation simply curbs the rate at which prices increase.

The final word in this chapter should go to former US president Richard Nixon, who might be the first politician to have invoked the *third derivative*.[6] During his campaign for a second term, Nixon announced that the rate of increase of inflation was decreasing. Take a moment to work through the three derivatives on display: (1) inflation itself (the rate of price increases), (2) the fact that inflation is increasing (the rate of the rate of price increases), and (3) Nixon's claim that the rate of increase (the rate of increases in the rate of the rate of price increases) is in decline.

Nixon's aim was probably intended to convey a picture of economic stability, which is not what the calculus of his statement amounts to (inflation was still rising, after all). The public can be forgiven for not seeing through the fog of his analysis. It may not be the scandal he will be remembered for, but it is a gross misappropriation of mathematics nonetheless – and one worth keeping in mind the next time a politician pledges to slash inflation.

Relations

The unexpectedness sameness of things

Few things in life promise the same certainty as mathematics. What, after all, could be more true than statements such as $1 + 1 = 2$?

In a 1939 essay, George Orwell warned that the slide towards authoritarian rule relies on casting aspersion on facts we would consider self-evident. The example he gives is an arithmetical one. 'It is quite possible', says Orwell, 'that we are descending into an age in which two plus two will make five when the Leader says so.'[1]

Orwell gives a starring role to the same sum in *1984*, where his protagonist Winston Smith wonders if the Inner Party, which exercises a vice-like grip on society's every thought, might declare that 'two plus two equals five'. The blatant lie might possess a certain validity, he reflects, if enough people are convinced by it.

The great Russian novelist Fyodor Dostoyevsky had a different take on presumed mathematical truths. He suggested that our deference to logic, which results in our automatic acceptance that two plus two makes four, signals a lack of intellectual agency. While he is prepared to accept four as the answer, he also wants us to consider an alternative:[2]

> I admit that twice two makes four is an excellent thing, but if we are to give everything its due, twice two makes five is sometimes a very charming thing, too.

Others have taken an even stronger stance, arguing that the expression $2 + 2 = 4$ comes apart in the real world. In his 1834 novel *Séraphita*, Honoré de Balzac noted:

> You will never find, in all Nature, two identical objects; in the natural order, therefore, two and two can never make four; to attain that result, we must combine units that are exactly alike, and you know that it is impossible to find two leaves alike on the same tree, or two identical individuals in the same species of tree.

The sum is invalid, in other words, because equality is an abstraction that is unsuited to physical objects. Even if we accept some notion of equality – two equal cups of tea, say, and two equal spoonfuls of sugar – it is not obvious how adding the two sugars to the two cups results in four of anything. Calculation, when wedded to context, can take on a whole new meaning.

These arguments rumble on in the digital age. Mathematics, for all its apparent objectivity, has a track record of going viral when people entrench themselves behind opposing viewpoints. The statistician Kareem Carr caused a Twitter storm in 2020 when he revived the debate around 2 + 2 = 5:[3]

> I don't know who needs to hear this, but if someone says '2 + 2 = 5', the correct response is 'What are your definitions and axioms?' not a rant about the decline of Western civilization.

Carr's tweet was a pre-emptive rejection of any suggestion that disputing 'obvious' mathematical statements is somehow an attempt to replace hard-fought truths with subjective opinions. It was inevitable that some of the responses to his tweet accused him of exactly that, with one commentator branding the claim a postmodernist effort to 'destabilize any sense of solidity and meaning'.[4] Yet most mathematicians agree with Carr that there is more to the equals symbol than meets the eye.

The symbol was first adopted by the Welsh physician and mathematician Robert Recorde in his sixteenth-century work *The Whetsone of Witte*; realising he was using a lot of ink saying that two things were the same, he introduced the shorthand '=' to 'avoid the tedious repetition', opting for a pair of parallel lines 'because no two things can be more equal'.

The problem is that using the '=' symbol does not necessarily mean that the two things on either side of it are the same in every respect (this is even true, despite Recorde's claim, of the two lines of the symbol itself).

Equality usually only gives a partial account of whether two things are the same. When, in 2015, a student in the United States lost marks for daring to suggest that 5×3 equals 15 because $5 + 5 + 5 = 15$, online commentators bemoaned a drop in educational standards.[5] According to the teacher's annotations, the student had calculated three lots of five (3×5) and should have instead reasoned that $3 + 3 + 3 + 3 + 3 = 15$.

In one view, the teacher has lost the plot. Surely it makes no difference whether we take three lots of five or five lots of three: you get 15 either way. In this sense, 5×3 really is the same as 3×5. But in another sense, they don't represent the same calculation. I could probably carry three boxes each containing five eggs home from the supermarket, but I'd struggle to manage with five boxes each containing three eggs. The difference between 5×3 and 3×5 is one of structure.

To most people, the expression $5 \times 3 = 15$ warrants no further scrutiny. To a mathematician, however, implicit within it is a notion of equality that requires precise definition. And in the same vein, outlandish statements such as $2 + 2 = 5$ may not be as absurd as they first appear.

This is not a philosophical argument any more than a sociopolitical one. It is simply the way mathematicians think about sameness: they demand to know the context, the *sense* in which two things are considered equal. This leads us to the broader notion of *equivalence*, which may be the creative licence Dostoevsky was seeking – and that Carr is inviting us to exercise – when studying how objects relate to each other.

Equivalence relations

I will not attempt to justify $2 + 2 = 5$ but will instead hang my hat on the following expression:

$$23 + 8 = 7$$

This is my variation of the absurd sum made possible, and it's one I swear by because my daily rituals are tied to it. I would guess that you have your own version of this equation, because the expression describes my sleep schedule. My nightly routine has me in bed by 11pm (or 2300 in military convention) and up by 0700 the next morning, ensuring I get eight hours of shut-eye. My children's sleep patterns tend to disrupt this routine with staggering predictability, but every night my alarm is set to the rhythm of that equation.

If numbers are our tools for making sense of the world, we must accept that they do not always behave as we would expect. When midnight strikes, instead of extending the counting sequence beyond 23, the hour resets to 0 and triggers another cycle.

Our decision to 'treat 24 as 0' results in a new and different-looking number system. In this 'modular arithmetic', you can perform calculations as normal, so long as you remember to erase any copies of 24. To compute 7×8, for example, you wouldn't leave the answer at 56. In a world where 24 counts for nothing, 56 is the same as 32, and 32 is the same as 8. In other words, this system yields the expression $7 \times 8 = 8$, as well as my strange-looking sum.

A mathematician would say these statements hold true 'modulo 24', where 'modulo' means 'set it to zero'. Mathematicians find the term useful in everyday conversation. When my maths professor friend David tells me he is 'All good modulo this darned paper I need to submit', he's saying that all would be well if he could somehow cast that paper aside.

Saying that two numbers are the same if they differ by some multiple of 24 is an example of what mathematicians call an *equivalence relation*, the point of which is to attach some notion of sameness to pairs of objects. We can relate pairs of objects in many ways, and we have an equivalence relation when the following three properties are satisfied:

- **Reflexivity** – every object under consideration relates to itself.
- **Symmetry** – if an object A relates to another object B, then B also relates to A.
- **Transitivity** – if A relates to B, and B relates to C, then A relates to C.

Blood siblings are one example; let's say for a moment that two people are only related if they share the same two parents. Our three conditions hold without any fuss:

- Everyone has the same parents as themselves (as weird as that sounds!).
- If I have the same parents as you, you have the same parents as me.
- If I have the same parents as you, and you have the same parents as Bob, I have the same parents as Bob.

So siblingness affords an equivalence relation, but the same is not true for friendships. For one thing, the relation fails the test of transitivity – if Alice is Bob's friend, and Bob is Charlie's friend, it does not follow that Alice and Charlie are friends. It may even fail the reflexivity test (unless you consider yourself among your own friends) and the symmetry test (in the case of unrequited friendship). Where equivalence relations are concerned, blood really does run thicker than water.

The point of an equivalence relation is to help us lump objects together so that we can treat them as a single group – an *equivalence class*. Whatever space of objects the relation is defined on, the equivalence classes form a neat partition. This means that every object belongs to precisely one equivalence class. In our time-telling example, we effectively reduced the entire collection of whole numbers to one of 24 types. There is an equivalence class containing 0 (it also contains 24, 48 and every other multiple of 24). There is another equivalence class containing 1 (as well as 25, 49 and every other number that leaves a remainder of 1 when divided by 24). There are also distinct equivalence classes for the whole numbers between 3 and 23.

Equivalence relations feature in many areas of maths. My favourite example, from geometry, will appeal to fans of cult video games such as *Asteroids*, where objects disappear from one end of the screen before popping up on the opposite side. The game designer has 'glued' together opposing points of the square. This can be thought of as an equivalence relation, where two points are 'related' if they are the same distance along opposite lines of a square. Grab a sheet of paper and try it: bring one pair of opposite edges together to get a cylinder, and then the remaining pair of edges to get a doughnut. The strange two-dimensional world of *Asteroids* was in fact the ordinary three-dimensional world of a doughnut (or a torus, to use its mathematical name). Equivalence relations alter the fundamental structure of things.

This way of thinking blurs our notions of what it means for two things to be equal. Mathematicians rely on precise definitions, but all of a sudden we're saying two things can be considered equal if they are somehow the same structurally (the term mathematicians like to use is *isomorphic*). This semantic ambiguity isn't something mathematicians can just sweep under the rug either. For several years a whole enterprise of 'automated theorem proving' has existed, where the idea is for computers to automatically generate mathematical results from statements that are fed to them. For this to work, however, the statements need to be free of ambiguity. Where mathematicians may be able to flit between competing definitions of equality, a computer demands a clear-cut definition. Some mathematicians have argued that equivalence, as interesting and useful a concept as it is, falls short of the requirements for equality.[6]

From a practical standpoint, equivalence classes serve as an organising mechanism. Why bother confronting every object individually when we can instead study a smaller number of equivalence classes?

Marketers use this ploy when segmenting customers into different types based on their buying habits or demographic. It may make sense for a bookseller to treat two middle-aged men with a predilection for crime thrillers as the same, serving them the same adverts and offers. Those two customers are not literally identical, but on the characteristics that the bookstore believes are most predictive of their next purchase, they may as well be indistinguishable.

An equivalence relation tells us what objects must have in common to be considered the same. It is an instruction to wilfully disregard all other aspects of their structure, a ploy we frequently use in the real world. In trading, equivalence goes by another name: *fungibility*. A fungible asset is one that cannot be distinguished from another asset of the same type. Paper currency fits this description: the £10 note in your wallet is, for all transactional purposes, worth the same as the one in mine. The notes are not literally identical, but we could switch notes without any material consequence. Fungible assets can be copied and exchanged: the value of stocks and bonds does not depend on what portfolio they belong to. Crude oil is fungible because one barrel is worth the same as the next. Non-fungible assets, meanwhile, dwell on the differences: the original *Mona Lisa* carries far more value than its countless replicas. In the digital space, non-fungible tokens (NFTs) apply this idea to cryptocurrency by assigning a unique signature – a piece of digital artwork, made up of specific snippets of code.

We routinely use equivalence classes in our social lives. In our everyday interactions, we constantly have to give thought to other people's likely beliefs and attitudes. Being able to place them in neatly defined categories is an indispensable shorthand when deciding which relationships to pursue, and which to avoid. When we form our social groups, we make value judgements on the characteristics in others that we think do – and do not – matter. We identify as members of certain groups based on highly specific attributes – our nationality, our religion, our political affiliation, even our sports team. In this context, an equivalence class is a select tribe of people that happen to share a particular characteristic. At play is a heavily simplified parsing and amplification of selected bits of our identities.

Our desire to belong to a group runs so deep that we will readily identify with other people on the flimsiest of terms. For instance, subjects in one study were assigned the personality trait of 'overestimators' and expressed favourable opinions of others who were given this label, even though it is void of any substantive meaning.[7] This drive towards tribal impulses takes root in infancy. In another study, this time involving three- and four-year-olds, the subjects were randomly assigned into one of two colour groups. Even without any meaningful knowledge of what the colours represented, the young subjects expressed stronger preferences for other children who had been allocated to the same group.[8]

Our shared markers, however superficial, become the basis for in-group favouritism – we think better of, and are more likely to help, those people we identify with. The danger here is that we create unnecessary distance from people outside of our chosen group. Social prejudices feed off the idea that we can get the measure of a person purely by dint of, say, their gender, ethnicity or skin colour. Here, an equivalence class serves as nothing more than a straitjacket that substitutes a person's individuality for a crass judgement based on a group they happen to belong to. One's own equivalence class, being familiar to its members, is deemed somehow superior to all those 'out-groups' it has been divorced from.

Ordering

If we cannot show two objects to be equal or equivalent, in whatever sense, we may resort to declaring one superior to the other. This is justified when comparing the size of numbers; we use the '<' symbol to denote that one number is 'less than' the another. '<' is an example of an *order relation*, a means of ranking items against one another. Humans like to compare things, but as we rank the world around us, we often find ourselves co-opting order relations in ways that may not be appropriate.

For me and my friends, the school playground was a place to taunt or be taunted, and the weaponry for our daily verbal conquests came from the latest sports results. For me, as a Manchester United fan, the football exchanges were all too easy – this was the late 1990s, where 'my team' won just about everything. Cricket was a different matter. My Pakistani heritage dictated that I stick it to my British Indian peers at every opportunity – easier said than done, given that Pakistan's first victory against their neighbours in major tournaments came in 2021 – by which time I had mostly grown out of tribalistic one-upmanship.

Instead, I opted for the next best thing. If I could identify a team that Pakistan had defeated who, in turn, had triumphed against India, then I reasoned to my sceptical peers that Pakistan's superiority over India was a logical consequence. My guiding principle was transitivity, which we have already encountered as one of the defining traits of an equivalence relation. It is also an inbuilt property of order relations, which is essential in any attempt at lining objects up against each other.

In my case, the relation centred on a notion of winning: 'If A beats B, and B beats C, then A beats C.' It was a convenient logic to take to the battleground of the school playground, albeit one that failed to live up to reality when Pakistan played India.

Transitivity certainly applies to the size of numbers: if I am taller than you, and you are taller than Ed, then I don't need the tape measure to infer my vertical superiority to Ed. Transitivity works in non-numerical contexts too – we've already seen this for siblings. But transitivity just as easily falls apart – replace siblings with cousins. Or take the following sequence of football results in 2023 that culminated in a low point for Manchester United fans.

Liverpool 2–5 Real Madrid
Real Madrid 0–1 Barcelona
Man Utd 2–1 Barcelona
Liverpool 7–0 Man Utd

My only crumb of comfort from that final result came from the mathematical realisation that it completed a non-transitive sequence. If the transitivity assumption held up, this sequence of results would have us conclude that Liverpool are better than themselves – a patent absurdity. The reason sport is so watchable (and maddening) is that it defies such predictive rigidity.

It is easy enough to manufacture a *non-transitive* chain. Take the following three dice, where the numbers repeat on opposite faces:

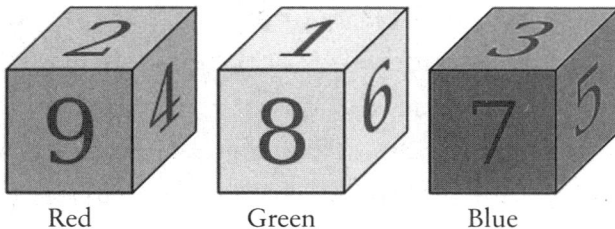

Red   Green   Blue

Which is the highest-scoring dice? They all have an average value of 5, so it is not immediately obvious. Let's instead compare the dice two at a time, one face at a time, starting with red and green. The possible combinations are:

Red	Green	Which is higher?
9	1	Red
9	6	Red
9	8	Red
4	1	Red
4	6	Green
4	8	Green
2	1	Red
2	6	Green
2	8	Green

Red wins out five times out of nine. Do the same with green and blue and you'll find that green wins out five times out of nine. You would think, then, that red will win out against blue, but blue actually defeats red five times out of nine. So red beats green, which beats blue, which beats red. Our effort to pit the dice against each other resulted in another cyclical, non-transitive chain that rather undermines the idea of ranking.

Non-transitivity can also rear its head in decision-making, resulting in all manner of conflicts. In particular, it defies the notion that humans are rational agents who always opt for logical choices. Even a choice as basic as what to have for dinner can result in non-transitivity. If my options for a takeaway one evening are Italian, Indian and Lebanese food, I might reason as follows:

- I fancy curry over pizza (Italian < Indian).
- But the curry is expensive, so I may as well go for the cheaper mezze option (Indian < Lebanese).
- But actually, a mezze is less appetising than a pizza (Lebanese < Italian).

Thus we end up in another loop: Italian < Indian < Lebanese < Italian. The reason is that we have applied different criteria in each comparison: in the first and third, we were guided by our palate, but in the second we were considering relative costs. The sum total of our preferential choices results in irrational behaviour. If you've ever faced decision paralysis at the supermarket, it is probably because you are finding separate reasons to buy each item on offer. If you went on cost – or taste – alone, your decisions would arrive more swiftly.

Non-transitivity might even explain our tendency to procrastinate.[9] If an article I'm writing is due on Friday, I can easily convince myself on the prior Monday that it's better to delay by a day – I might have more energy then, or a greater incentive to complete the article with the deadline looming. When Tuesday comes around, of course, I exercise the same rationale and delay to Wednesday, and then to Thursday. But on Thursday I'm in panic mode and wish I'd started on Monday after all. I've been caught in another decision cycle: Monday < Tuesday < Wednesday < Thursday < Monday. Only an extension to my deadline or a highly productive Thursday evening will get me through the task. In the future, the only way to break the procrastination cycle is to avoid it altogether: get started early to avoid the panic and regret of leaving things too late.

We have non-transitivity to thank for childhood games such as rock–paper–scissors, not to mention a flurry of internet memes that show how assuming transitivity can lead to preposterous outcomes. My favourite among them: *humans eat cows, and cows eat grass ... so humans eat grass.*

Transitivity is appropriate when there's a meaningful and consistent way to rank items, but it is very rare that we can neatly stack our options against each other; more typically, we must adopt multiple criteria that result in conflicting preferences.

This is a good moment to mention an oft-cited result from mathematics that brings the usefulness of transitivity into question. In its original framing, it is known as the secretary problem:

You want to hire a secretary from a pool of some fixed number of candidates. You interview candidates in turn and after each one make an immediate decision to either hire them (at which point the process terminates), or to reject them and move on to the next candidate. What strategy maximises your chances of finding the best candidate?

The standing assumption in this problem is that the candidates can be ranked. The optimal strategy has been shown to be:

- Interview the first 37 per cent of candidates and reject them all.
- Among the remaining candidates, if you interview a candidate that is better than all the rejected candidates, hire them and terminate the process.
- If no such candidate emerges, you must hire the final candidate.

The method may seem crude, yet it results in the best candidate being hired a remarkable 37 per cent of the time.* If you subscribe to the assumptions of the

---

* The repeat occurrence of 37 per cent is no coincidence – the problem is underpinned by the appearance of Euler's number e, which is around 2.7, and $1/e$ is approximately 0.37.

problem, the strategy will guide your decision on whether to keep searching for the perfect hire or to settle for the best candidate you have seen so far. You could even apply the method to your romantic conquests, but the predetermined rejection of 37 per cent of your potential mates may raise eyebrows (not to mention some serious ethical questions).

But what of those assumptions? They are replete with issues: is it ever appropriate to accept or reject candidates on the spot, and what if we don't know the number of potential candidates in advance? Another assumption is that the candidates can be ranked in a list that orders them according to their measure of suitability for the role. This is problematic: for one thing, your role is likely to have multiple criteria, and most candidates will excel in some of those areas but not others. When weighing up one strong candidate over another, you are surely better off making judgements that seek to understand the trade-offs between what each brings to the role. The final decision may always be binary, but the idea that this can be determined through a ranking mechanism seems crass, unless the basis of their hiring is an objective measure of performance.

Let's indulge the assumption for a moment and say you are confident in your ranking procedure. Now suppose you are one member of a recruitment panel, alongside two colleagues, Deedie and Elias. For your next role, you are presented with three candidates: Alice,

Bob and Charlie. After the interviews, you each proceed to rank the candidates, confident in your methodology. The results are:

	1st	2nd	3rd
You	Alice	Bob	Charlie
Deedie	Bob	Charlie	Alice
Elias	Charlie	Alice	Bob

There is a case to be made for Alice, and as her staunchest advocate you are the one who will make it. Here's what you might say:

- Both you and Elias have placed Alice ahead of Bob, so Alice deserves the role ahead of Bob.
- Now compare Bob and Charlie: both you and Deedie have placed Bob ahead of Charlie, so Bob deserves the role over Charlie.
- Surely, then, Alice deserves the role ahead of Bob and Charlie, which means she wins the contest overall.

Can you see the problem? All three candidates have performed in the same way overall, each earning a first, second and third place. There's no way of distinguishing one from another – and we could have argued just the same for Charlie and Bob.

Your attempt to secure Alice's offer relies on a faulty assumption of transitivity ('Alice outperforms Bob, and Bob outperforms Charlie, so Alice outperforms Charlie'). Compare Alice's votes with Charlie's: only you have placed Alice ahead of Charlie, so Charlie wins that particular duel. We're left with a deadlock, owing to the non-transitive cycle: Alice > Bob > Charlie > Alice.

This 'voting paradox' demonstrates that even when individual preferences are transitive, a group's aggregate preference may not be. It also shows that transitivity – and thus a meaningful ordering – is hard to come by. A theorem of Kenneth Arrow (which contributed to him winning the Nobel Prize for Economics in 1972) lists three requirements of a fair voting system:

1. Transitivity: if the system ranks candidate X above candidate Y, and Y above candidate Z, then it must rank X above Z (so the system avoids endless cycles that loop back on themselves).
2. If everyone votes for candidate X over candidate Y, then the system ranks X over Y.
3. The final ranking of candidate X over candidate Y should not depend on how voters ranked a third candidate Z.

In our example, we used a particular method to select our candidate, *majority rule*, which says that candidate

X is preferred to candidate Y precisely when a majority of voters opted for X ahead of Y. The system adheres to the latter two conditions but, as we've seen, it violates the first.

There are any number of alternative voting methods we could employ. The Borda counting method assigns each candidate points corresponding to their rank (specifically, each voter is assigned points equal to the number of candidates placed below them). This innovation does not violate transitivity because Alice, Bob and Charlie each score three points and are therefore not ranked above one another, and there is no risk of the candidates looping back on themselves. But suppose a fourth candidate, Frank, comes into the mix, and you rank him thus:

	1st	2nd	3rd	4th
You	Alice	Frank	Bob	Charlie
Deedie	Bob	Charlie	Alice	Frank
Elias	Charlie	Frank	Alice	Bob

Now Alice scores five points, Bob scores four points and Charlie scores five points. The addition of Frank has changed our ranking system: Bob has been knocked down a peg, violating the third of Arrow's conditions.

What Arrow showed was mildly depressing: in any election where individual voters rank candidates and there are at least three candidates, the only scenario in which all three conditions are met is the one where the preferences of the group precisely reflect the preferences of one particular individual. We have a word for this: *dictatorship*. The thought may have crossed your mind that, in the above scenario, your own judgement should count for more than that of Deedie or Elias. Arrow's so-called 'Impossibility Theorem' lends you some justification, confirming that only a dictator can safeguard all three conditions for a fair election.

There is some reprieve, however; in larger populations, the chance of a non-transitive cycle has been shown to be negligible.[10] Defenders of democracy can breathe easy, but in smaller societies we should remember the elusiveness of transitivity – and the compromises forced on us when we insist on ranking candidates. Humans are multifaceted; we should avoid attempts at placing ourselves in a neatly ordered list.

The median

It pains me to see mathematical terminology being misapplied. One word that pops up frequently is *median*; political commentators may refer to the *median voter*, educators the *median student*.

As a mathematical concept, the median is straightforward. It is a type of average that corresponds to the middle item of a list. Line up a group of five friends, shortest to tallest, and the person in the middle – the third person along – can be said to have the median height within the group (with an even number of objects, we take the halfway point of the middle two objects).

As an alternative to the mean, the median has its uses. If Elon Musk were to turn up at your local restaurant one evening, the mean wealth of the diners would surge into the millions. While it is strictly true that the *average* diner would be a millionaire, this says more about one person's ability to distort the overall picture of wealth than it does the diners' prosperity.

If we instead rank everyone in the room by wealth, Musk will jump straight to the front of the line, but his presence affects the median wealth only a little – the middle-ranked diner would move along by one place. Thus the median is useful precisely because it is resilient to the effect of outliers.

But in invoking the median, there is a supposition that objects can be ranked. For things that can be quantified – such as people's height or wealth – this is a safe assumption, but in other cases the median assumes too much. When educators speak of the 'median student', it is not the students per se that they are ranking, but their academic performance (usually their scores on standardised tests). The median only makes sense if there is a quantifiable metric in play, which usually represents an oversimplification. Once we embrace students for their individuality, we realise there is no meaningful way to rank them. As for the 'median voter', where do we even begin? We might categorise voters according to their political beliefs, but what could it possibly mean to rank them?

When 'median' is used in everyday parlance, there's a reasonable chance it is a lazy substitute for the word 'typical' (itself an imperfect term, but one that at least has broad interpretability). Mathematical language can make us sound clever, but it demands precision; without due attention to the details, our words may fail to convey anything meaningful at all.

4

Sets

Counting from zero to infinity – and beyond

In accounts of post-apocalyptic survival, mathematics features very little. *The Knowledge*, Lewis Dartnell's manual for how we might reboot civilisation in the event of cataclysm, offers a first-principles approach to reviving the everyday systems we rely on: agriculture, medicine, energy production and so on.[1] There is no explicit mention of mathematics, which, given the obvious importance of numbers to commerce and governance, is a surprising omission.

But numbers would not be the only lost treasure of mathematics. Let's suppose we have licence to add one more chapter to the survival guide, devoted solely to the aim of reviving mathematics – not just the numbers we would need for double-entry bookkeeping, but shapes, high-dimensional vector spaces, logic and everything else maths has to offer. To have any hope of recovering such a breadth of concepts, we have to get down to the brass tacks of the subject and identify the most fundamental object of study.

One good candidate is the *set*, which is nothing more than a collection of objects (or *elements*). Those elements can be anything at all: numbers, shapes, people, colours. They may even be sets themselves. Sets can be finite or infinite. We can even have a set containing nothing at all (the so-called 'empty set', denoted ∅).

{all positive numbers}

{red, blue, yellow}

{apple, Paris, {2, 3, 5}}

Three examples of sets, using the standard { } notation.
The third set contains three elements, the third of which
is itself a set containing three elements.

The definition is deliberately simple. We are merely laying foundations, which we hope will lead to familiar pastures such as the number system. Here is a flavour of how the counting numbers can be described with sets:

- Define 0 to be ∅, the empty set. This set contains no objects but it is itself an object – this is a Big Bang moment for mathematics; from literally nothing we have conceived of a tangible entity, a set, which happens to contain nothing in it.
- Now put this object, 0, into a brand-new set. This new set, which we can write as {0}, contains

a single object (namely, the set that itself contains nothing!).

- By now we have two objects, which we have called 0 and 1. We are ready to define our next object, which we'll call 2, as the set containing these two objects – that is, $2 = \{0, 1\}$.
- You may see where this is going: define 3 to be $\{0, 1, 2\}$, the set containing the three previous sets. And so on.

You might recall visualising sets with a Venn diagram, a mental model that proves so useful that a section at the end of this chapter is devoted to it. The vision of early pioneers of set theory was akin to a unified theory of mathematics. Their aim was to express vast areas of maths in this language. Sets fall short of delivering the perfectly logical system mathematicians strived for – in the next chapter we'll see why. But as an object of study sets offer their fair share of useful concepts. This chapter will look at some of the ways in which they can add to our arsenal of thinking tools.

Disclaimer: if you are an older American reader who was subjected to the New Math curriculum at school, do not fret. New Math was rooted in the idea of maths as a formal, abstract subject. The approach was championed by US policymakers who, reacting to Russia's launch of the artificial satellite Sputnik in 1957, felt that young students of maths needed a more rigorous grounding in the subject to compete with their international peers. New Math placed *set theory* at its heart, taking inspiration from the idea of the set as the most foundational object of study. But things did not go as planned, for the obvious reason that this way of teaching maths proved too baffling for schoolchildren, not to mention their parents and even their teachers. This chapter will not take its cues from New Math; our aim is to strip set theory down to its most essential ideas and apply them to our everyday thinking.

Cardinality

Because a set is defined as a collection of objects, it is natural to ask how large this collection is. This is known as its *cardinality* and, for finite sets, it works exactly as one might expect: the set of primary colours has a cardinality of three, the set of holes on a golf course has a cardinality of eighteen and the set of stars in the observable universe has a cardinality of around 10^{24}. The empty set, containing nothing at all, has a cardinality of zero.

Cardinality captures a notion of quantity in a manner that is independent of specific objects. The number 3 may be understood as the property that is common to the set of primary colours, the set containing the blind mice in the popular nursery rhyme, and every other set that contains that many objects.

This might feel a bit chicken-and-egg. After all, how can we describe two sets as having the 'same amount' of objects without first having a language for numbers? How, for that matter, can we describe one set as larger than another? The answer lies in pairing. Imagine we have a bowl of apples and another of oranges, and we want to determine which we have more of. Instead of counting the number of apples and oranges in turn, we can instead remove one of each at a time, pairing them until no items remain in one of the bowls. One of two things will happen: either there is no fruit left over, in which case the apples and oranges are in perfect correspondence, and we can conclude that the two bowls have the same cardinality. Or there will be some objects left in one of the sets – perhaps a few spare oranges – in which case, the bowl of oranges has a higher cardinality.

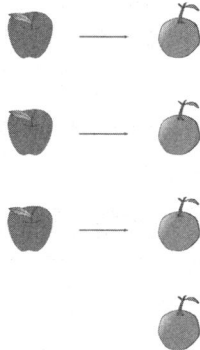

By pairing apples and oranges as far as we can, and seeing that there is an orange left over, we can conclude there are more oranges – without having to count the number of fruit.

The pairing mechanism – known as Hume's Principle, after the philosopher David Hume – suggests that our notions of relative size are independent of any specific counting scheme. It is how our ancestors were able to keep track of their flocks, and many tribespeople still use it today. By denoting each of their sheep with a tally mark, they can quickly tell when one has gone astray by pairing the sheep with the marks and checking if any marks are left over.

Most pairing schemes involving finite collections of objects are straightforward enough. If I pair my socks one at a time and find myself holding a single sock at the end, I know that there will be another sock unaccounted for (of course, many more pairs may also be missing). Or if I place every knife and a corresponding fork back in the cutlery drawer and find myself wielding a lonely knife, I must go searching for a fork.

For a less obvious pairing scheme, let's return to our fruits and add a few more to the mix: apple, orange, banana, mango, pear. From these five fruits, we can select any number and make a fruit salad (assume that when a fruit is chosen, only one is used in the salad). How many possible fruit salads can we make? This is a problem of combinations, which is a focus of Chapter 7. For now, you will surely agree that the answer does not leap out. There are so many options, after all, such as:

- orange–banana–mango–pear
- apple–banana–pear
- orange–pear
- pear (lone fruits do not strictly constitute a fruit salad, but let's count them among the options)

In the language of this chapter, we started with a set of five objects (apple, orange, banana, mango, pear), and we want to know how many *subsets* there are whose objects are drawn from the original set. If we put all these selections (or subsets) together, we end up with a whole new set called the *power set*. Our aim is to determine the cardinality of this new, larger set.

The answer can be found with an elegant pairing scheme. The idea is to turn every possible fruit salad into a five-digit string comprising solely 0s and 1s. Here's how to do it. Given a fruit salad, check through the five fruits in order: apple, orange, banana, mango, pear. One by one, ask if each fruit belongs in the salad. If it does, write down a 1. If it doesn't, write down a 0. By the end of the process, you will have a five-digit string containing 0s and 1s. Our salads above would correspond to the following strings:

- orange–banana–mango–pear   01111
- apple–banana–pear   10101
- orange–pear   01001
- pear   00001

The point is that, for any given salad, the five-digit string is unique. What's more, every five-digit string corresponds to one of our salads (we have to allow for the slightly odd case where we forgo all five fruits – this trivial selection corresponds to the string 00000).

The hard work is done: we've paired every one of our fruit salads with one of those five-digit strings. In mathematical parlance, we have found a *bijection* between the set of every fruit salad and the set containing all the five-digit strings containing just 0s and 1s. So asking how many fruit salads we can make is equivalent to asking how many five-digit strings there are. The latter turns out to be straightforward because for each of the five digits, we have two options: 0 or 1. Each digit therefore doubles the number of possibilities. This gives a total of $2 \times 2 \times 2 \times 2 \times 2 = 32$ five-digit strings, and therefore 32 possible fruit salads.

To summarise, we started with a set of cardinality 5 and, through a clever pairing scheme, we worked out that its power set (the set of all the combinations of objects from the original set) has a cardinality of $2 \times 2 \times 2 \times 2 \times 2$ (or 2^5 for shorthand, where 5 denotes how many times the 2 occurs in the product). There is nothing special about fruits, or the number of objects we started with. So we can generalise and say that if a finite set has cardinality N (where N is a whole number), its power set will have cardinality 2^N (2 multiplied by itself N times). That is a combinatorial explosion of possibilities.

We have now laid the groundwork for approaching the cardinality of *infinite* sets, where things take a surprising turn.

An infinite tower of infinities

If we have two collections of infinite objects, we cannot hope to count either in full. We also know, from Chapter 1, that it's not enough to declare every infinite set as having the same size – there are at least two levels of infinity. There is the *countable* infinity, which applies to sets where the objects can be listed, the most obvious example being the natural numbers. But there is also the *uncountable* infinity; the set of real numbers – by which we mean all the numbers on the number line – turns out not to be listable in any way.

We can understand this difference by invoking the idea of pairing schemes once again. Suppose we have two sets – let's call them A and B – and that we can find a pairing scheme that serves up a unique object in B for every object in A. The cardinality of B is then *at least as large* as the cardinality of A. Now, if there is any pairing mechanism that matches every element of A with every element of B (a bijection), then the two sets have the same cardinality. But if every conceivable pairing scheme leaves behind some objects in B, then B has a strictly larger cardinality than A. This is exactly how we handled the case of finite sets (apples versus oranges and fruit salads versus five-digit strings) but, crucially, it works just as well for infinite ones.

Another way of saying a set is countable is that its objects can be paired off with the natural numbers (1, 2, 3, ...). When we demonstrated that the fractions are countable in Chapter 1, we did so in exactly this way, showing that they have the same cardinality as the natural numbers. But we also found that any attempt at pairing the natural numbers with the real numbers is doomed to failure: there are so many more of the latter that no attempt will get them all; there will always be leftovers. So the real numbers have a higher cardinality than the natural numbers.

Are there other levels of infinity? Yes, infinitely many. The mathematician who first distinguished between the countable and uncountable levels, Georg Cantor, devised a scheme for creating even larger infinite sets. It is basically the same scheme we used to count the number of fruit salads. Cantor realised that power sets are much larger than their original sets (for finite sets, remember, this is on the order of exponential growth). His key insight, the details of which are a little too technical to include here, is that the same is true for infinite sets – that is, the cardinality of an infinite set is less than the cardinality of its power set. Whatever set we start with, considering all possible combinations

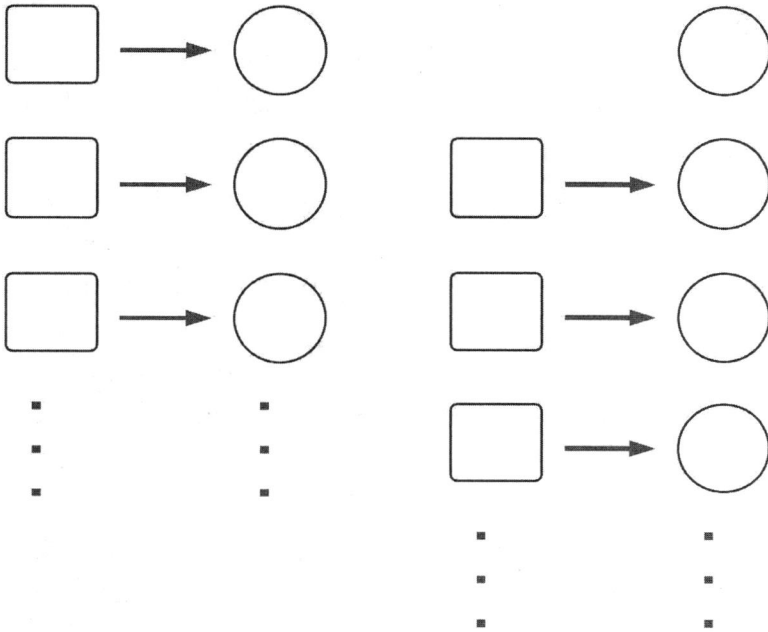

Here we have two sets, the elements of which are denoted by squares and circles. On the left, we have a pairing that matches every element of each set. In this case, the two sets have the same cardinality. On the right, we have an attempted pairing mechanism that matches every element of the first set but misses out elements of the second set. If every attempted pairing fails like this, then the second set has a larger cardinality.

of objects in that set gives rise to another, distinctly larger set.

The incredible consequence of Cantor's scheme is that it gives rise to an infinite tower of infinities, each with a cardinality larger than the previous one. Starting with the most basic 'countable' infinity of the natural numbers, the power set has a higher cardinality.* To jump to an even higher infinity, we take the power set of our new set, which is the set of all the combinations of all the combinations of the natural numbers. And the power set of this set (the combinations of the combinations of the combinations) has a larger cardinality still. By now our sets are layered in abstraction and difficult to keep track of mentally. But the key point is that we could keep the process ticking forever, generating a new level of infinity each time

Cantor was chastised for his attempts to bring infinity into mathematics; in doing so, he had shattered the conventional view of the infinite as a unitary, monolithic idea. The notion of a chain of infinities grated with several of his contemporaries – Leopold Kronecker branded him a 'scientific charlatan' and a 'corrupter of youth'. Cantor suffered from severe bouts of depression (while the exact causes are unknown, this hostility could not have helped).

But Cantor's critics were short-sighted. Firstly, his revelation of infinity as a pluralistic concept is not as

---

* In fact, it turns out that the power set of the natural numbers can be paired off with the real numbers. So that distinction between countable and uncountable infinities in Chapter 1 is just the first link in Cantor's infinite chain.

Level of infinity	Three examples of elements in this set
'Countable' – the natural numbers	1, 2, 3
Power set of the natural numbers – combinations of natural numbers	{1, 2}, {2, 5, 7}, {5, 9, 10, 11}
Power set of the power set of the natural numbers – combinations of combinations of natural numbers	{{1, 2}, {2, 5, 7}}, {{11, 18}, {40, 41, 42, 43}}, {{1, 2, 3}, {4, 5, 6}, {7, 8, 9}}

The first three levels of infinity in Cantor's infinite tower

far-fetched as it may seem. Just think of the love you hold for your nearest and dearest. Each of those relationships is, no doubt, a blessing that defies any finite description. But it is unlikely that your love for those people exists in equal, infinite measure. It is possible to have relationships that seem to us, at the time, unsurpassable in their significance – a bond with a soulmate, or with parents or a sibling – and for these relationships to subsequently be disrupted by a form of love that seems even more impossibly infinite, such as the love parents experience for their newborn children.

What Cantor really showed is that the more we peer into ethereal concepts like infinity, the more they reveal themselves in all their glorious complexity. That there is no end to his tower is most befitting of infinity's status as an unattainable ideal.

After that foray into the infinite, let's return to the more grounded case of finite sets for another novel application of cardinality.

Pigeonhole principle

I learned about the pigeonhole principle two weeks into my undergraduate maths degree. It was the early noughties, when it was possible to get through an entire term without checking one's email. Communication between peers relied largely on text messages, and physical notes that were left in our mailboxes – quaintly known as pigeonholes.

The pigeonhole principle is one of those ideas in mathematics that is blindingly obvious at first but, upon closer inspection, can.lead to surprising results. It says, very simply, that if you try to place pigeons into pigeonholes and there are more pigeons than holes, then at least one of the pigeonholes must accommodate more than one pigeon.* In terms of our pairing mechanism from earlier in the chapter, the principle merely observes that if one set is larger than another, it is not possible to pair

---

* The original formulation used the analogy of drawers rather than dovecotes.

off every object of the larger set with a different object in the second set.

I did say it was obvious! It doesn't just work for pigeons, of course: as long as you know that a collection of objects (in this case the pigeons) outnumbers a range of possible values (the pigeonholes), the principle kicks in. It's how you know, for instance, that in any five-card hand from a standard deck of cards, at least two will share the same suit. Or that when a chair is removed during a round of musical chairs, one participant will fail to find a seat.

The principle can also help demonstrate not so immediately obvious facts, such as the following:

> *In London, there are two people who have the same number of hairs on their head.*

If we allow bald people, the claim is trivial, so let us exclude them from the picture. You might resort to visiting the capital and interrogating its citizens one scalp at a time, tallying up each person's hair count. But to preserve your sanity (and theirs), you might instead apply the pigeonhole principle by invoking the following facts:

- The population of London is around nine million.
- The average human head contains 100,000 hairs.

The pigeons are the millions of non-bald Londoners, and the pigeonholes are the possible hair counts,
between 1 and, let's say, 200,000. There simply aren't
enough numerical options for each Londoner to possess
a different quantity of hair. A few bushy-haired outliers
aside, the vast majority of Londoners will have somewhere between 1 and 200,000 hairs. Even if you could
find someone with a single hair, someone else with two
hairs, yet another person with three hairs and so on to
200,000, you could not avoid two individuals sharing a
hair count.

We may lack precise knowledge of the quantities
involved – the estimates above, for instance, were sourced
from a Google search – but by dint of very basic logic,
we are able to solve counting problems without doing
any counting.

The pigeonhole principle rears its head in unexpected
places, and even at parties. At any gathering, you can
be sure that two of the partygoers will know the same
number of people in the room. In this case, the pigeons
are the partygoers while the pigeonholes correspond to
the number of friends they might have in the room. The
argument is as follows:

- Suppose there are 100 people at the gathering
  (though the same argument will apply for any
  number).

- First, consider the case where someone (maybe you) knows nobody.
  - » If there is another loner in the room, the two loners share the same number of acquaintances.
  - » Otherwise, the loner is unique, which means the remaining 99 people at the gathering have between 1 and 98 acquaintances. Now bring in the pigeonhole principle: if there are 99 people and 98 choices at least two people must share the same number of acquaintances.
- In the case where there are no loners, we have 100 people with 99 choices (they can have between 1 and 99 acquaintances). The pigeonhole principle takes effect once again, and two people must have the same number of friends.

These examples veer towards the 'interesting but useless' end of the spectrum, but the pigeonhole principle also lends itself to practical applications. It is called on, for instance, to show that every compression algorithm that reduces the size of at least one file must also accept some loss of data. The argument goes as follows:

- Suppose we have a compression algorithm that can reduce the size of at least one file. Consider the smallest file that the algorithm is able to compress into an even smaller file.

- This smaller, compressed file has a particular size. Let's say there are N files of this size. When these N files are run through the compression algorithm, their size is not affected (i.e. they do not become any smaller) because they are smaller than the smallest file the algorithm is able to compress.
- Altogether we have found $N + 1$ files that are compressed into this smaller size: the original file in the first step, along with the N files in the second step. Each of the $N + 1$ files must be compressed into one of the N files of the smaller size.
- Now we apply the pigeonhole principle: two of the $N + 1$ files must compress into the exact same file. This compressed file cannot be decompressed reliably, because we won't know which of the two files it came from. It is in this sense that the compression algorithm 'loses' some data.

A more general version of the pigeonhole principle was formulated by the computer scientist Edsger Dijkstra. Like the original, it might seem obvious at first, but it has surprising applications. Djikstra's version says that the highest number of pigeons in any single box will be at least equal to the average number of pigeons in all the boxes. Imagine this time that we have 100 pigeons stuffed into 10 pigeonholes. The pigeonholes accommodate an average of 10 pigeons. Some may contain more, others less, and the result says that there must be a pigeonhole with at least 10 pigeons – if all the pigeonholes contained fewer than 10 pigeons, the total number would be shy of 100.

We'll use Dijkstra's variant to show that, across the world, at least forty-two people share the same gender, number of children, three-letter initials and birthday. We shall assume the following:

- The number of children any person has ranges from 0 to 9 (i.e. there are ten choices).
- Gender options are male, female or other (three choices).
- The number of three-lettered initials is 26 × 26 × 26 = 17,576.
- There are 366 possible birthdays (accounting for leap years).
- The world's population is roughly eight billion.

The total number of combinations of the four characteristics is

$$10 \times 3 \times 17{,}576 \times 366 = 192{,}984{,}480$$

The pigeons are the eight billion people, and the pigeon-holes are the 192,984,480 combined characteristics that they can fall into. The basic version of the pigeonhole principle says that at least two people will share the same characteristics (since eight billion is greater than 192,984,480), but Dijkstra's version goes further. Since the average number of people sharing these combined characteristics is given by

$$\frac{8 \text{ billion}}{192,984,480} = 41.45$$

we can infer that one of those 192,984,480 combinations of characteristics is guaranteed to be shared by at least forty-two people. Humans have more in common than we sometimes appreciate – even as the population of our species has exploded into the billions, we cannot escape collisions between our various descriptors. The pigeon-hole principle (and its variants) can put a number on the things we have in common.

Venn diagrams

The popularisation of Venn diagrams may be the one positive vestige of the New Math curriculum, and it's now a go-to model for purveyors of viral social media content. The diagram, named after the mathematician, logician and philosopher John Venn (1834–1923), is a means of showing how different sets of objects or ideas relate to each other. In a Venn diagram, each set is represented as a circle, and the diagram visualises the ways in which they may intersect.

The two-circle Venn diagram reminds us that two seemingly disparate categories can overlap, forcing us to notice the intersection of things that we might otherwise consider disconnected. The boss at an education company I once worked for would urge us to 'remember the Venn diagram' whenever it was suggested that two objectives existed in tension with one another. A recurring example was the need to generate profit while at the same time delivering on the company's social mission to raise learning standards. The best way to achieve long-term social impact, he would tell us, is through a sustainable business model. This framing situated us squarely at the intersection of both capitalist and socialist ideals, with no room for 'either/or'.

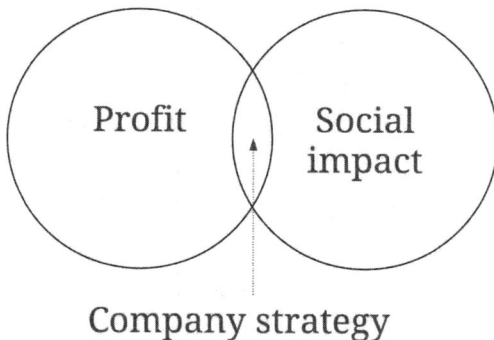

Company strategy

In the arts, the creative impetus to explore the intersection of different styles gives rise to whole new genres. When Miles Davis plugged his trumpet into effect pedals more commonly used by rock guitarists, he was criticised by jazz traditionalists for 'betraying' their genre. Other critics welcomed the blend of styles, and 'jazz fusion' was born. If you take any two music genres, there's a reasonable chance a hybrid has been attempted somewhere, from 'hip house' or 'rap metal' to 'blues rock'.

A common corporate marketing ploy is to place a company at the intersection of a two-category Venn diagram. Here is one I've used in pitches to publishers, in an effort to persuade them that there's an audience for my writing. It suggests that while you will find a plethora of books that popularise maths and a fair few in the smart thinking genre, my books are among the rare breed that capture the best of both. Convinced?

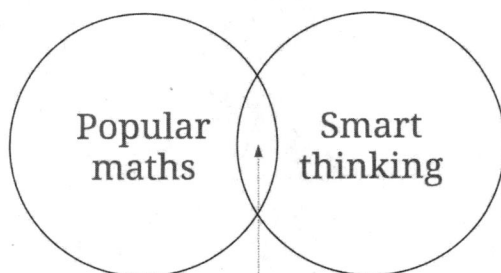

Popular maths — Smart thinking

My books?

Adding a third circle gives rise to more combinations; more opportunities to distinguish one idea or object from another. The centre of the Venn diagram is the sweet spot that captures the very best of everything, as is literally the case in the following example.

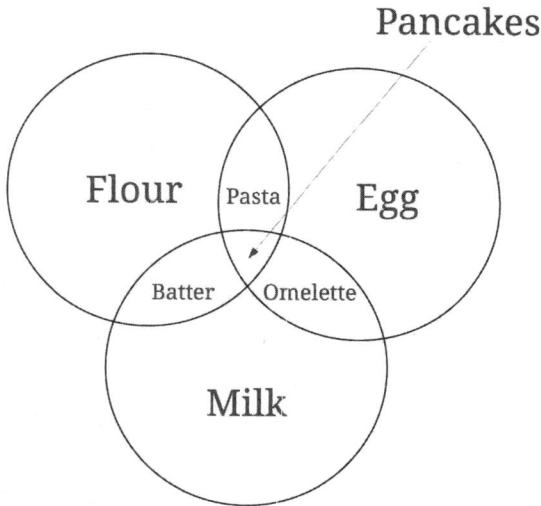

Pancakes

Flour   Pasta   Egg

Batter   Omelette

Milk

The intersection represents more than the sum of individual parts; consuming flour, egg and milk in turn does not, after all, sate one's appetite in quite the same way as pancakes.

This applies to people too. Venn diagrams are also a way of understanding the relationship between individual social identities and broader systems of power in our society. The American civil rights activist and scholar Kimberlé Crenshaw coined the term *intersectionality* to convey how demographics such as race, class and gender combine to privilege some groups over others. She cites the example of a group of five Black female auto workers who were laid off by General Motors in the 1970s when the company introduced a seniority policy that meant less experienced staff were the first to be let go. Because the company had only started to hire Black women in 1964, the policy had a uniquely damaging effect on this group.[2] Yet the legal case for unfair dismissal failed because in the court's view, the fact that General Motors employed African American male factory workers showed it did not have a race problem, and its employment of white female office workers, meanwhile, suggested women were also treated fairly. As Crenshaw explains, the judge refused to consider the claims of racial and gender discrimination in tandem, insisting on treating the plaintiffs as either Black or women. Intersectionality has become a lightning rod in America's ongoing culture wars, despite its relatively straightforward premise that individuals belonging to multiple minority groups face unique forms of discrimination due to overlapping systems of oppression.

Euler diagrams and other variations

Venn diagrams go out of their way to highlight every possible intersection, but they are limited to three sets: it is a mathematical impossibility to draw a Venn diagram with four categories. If we remove the arbitrary constraint that they have to be circles, we are able to extend the notion to four categories or more, thanks to a construction of geneticist Anthony William Fairbank Edwards. These Edwards–Venn diagrams work by segmenting the surface of a sphere. Here is how it renders for four categories, where every possible intersection is still accounted for.[3]

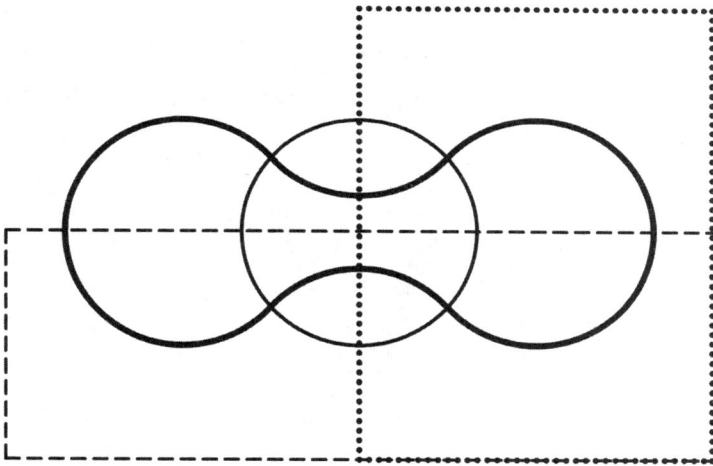

Venn would be pleased – he proposed his own version of a four-category diagram using ellipses[4] – but the dizzying array of combinations may bring the usefulness of these higher-order diagrams into question.

Venn's defining innovation was the requirement that every intersection is displayed. Many before him devised similar constructions that were free of this restriction;

one of the earliest examples comes from the Valencian scholar Juan Luis Vives, who in 1555 used triangles to visualise the syllogisms of Aristotle.[5] Here, for instance, is an illustration of the syllogism:

All A is B
All C is A
Therefore, all C is B

(The well-trodden statement, 'All men are mortal, Socrates is a man, therefore Socrates is mortal' is an example of this syllogism.)

The thirteenth-century philosopher Ramón Llull used logic diagrams as an attempt to win over Muslims to his Christian beliefs. In his work *Ars Magna*, he suggested that we can obtain knowledge of all things by exploring possible combinations of their basic principles.[6] To understand lofty concepts such as how the mind works or its relationship with God, one would just combine simpler, more fundamental concepts. Here we see the words 'Esse', 'Verum' and 'Bonum' in a Venn-like diagram, denoting the idea that 'nothing Exists which does not possess Unity, Truth and Goodness'.

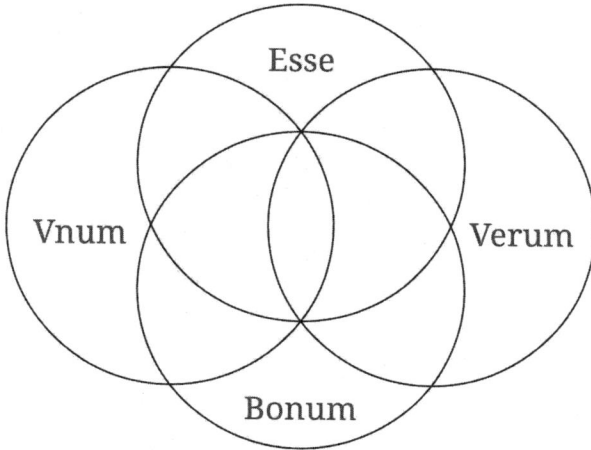

Llull's visualisation inspired the seventeenth-century polymath Gottfried Leibniz, who was probably the first person to use these diagrams to study logical propositions. Leibniz opted for overlapping circles, as well as overlapping lines, to denote how two statements may relate to each other. Each portion of the diagram corresponds to some statement of truth; the intersection denotes that 'Both statements are true', whereas the portion to the left signifies that 'Statement B is true but A is false' and the portion to the right signifies 'A is true but B is false'. The statement 'Neither A or B is true' is covered by the portion lying outside of the two circles (or two lines), meaning that all possibilities are accounted for. This work foreshadowed the way that search engines facilitate online shopping. If I head to amazon.co.uk and search for 'bestselling maths books', far from searching through every item in its store, Amazon will split up its data: there will be one collection of maths books and another collection of bestsellers. It then finds the intersection of those two collections to serve up the items I'm after.

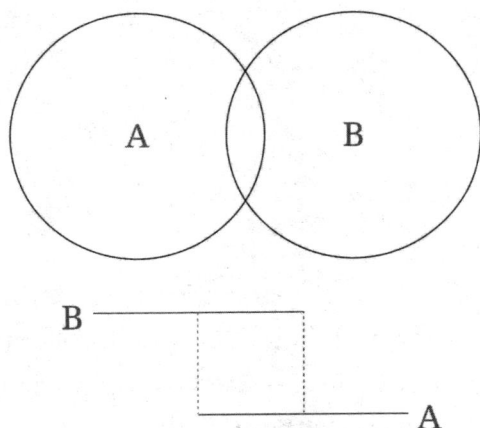

In the first diagram, 'B is true but A is false' is captured by the portion of the left-hand circle that does not overlap with the right-hand circle. In the second diagram, the same statement is captured by the portion of the upper line that is to the left of the first dotted line. Similarly, 'A is true and B is false' is captured by the portion of the right-hand circle that does not overlap with the left-hand circle, and by the portion of the lower line that is to the right of the second dotted line. Finally, 'A and B are true' is captured by the intersection of the two circles, and by the portion of the upper and lower lines between the two dotted lines.

Almost a century after Leibniz, the Swiss mathematician Leonhard Euler adopted a near-identical approach. In an Euler diagram, the circles need not intersect. At the other extreme, you may find one circle entirely contained in another. In the following diagram, A might represent citizens of the United Kingdom and B might be citizens of Europe (yes, the UK remains a part of Europe), while C denotes citizens of a different region altogether, say Asia.

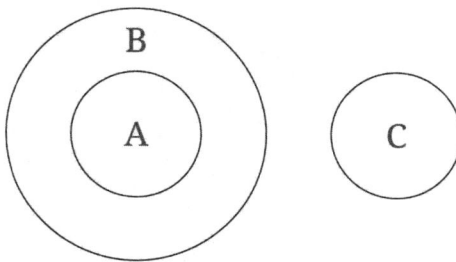

It was about a century later that John Venn iterated the concept again while studying the systems of logic developed by George Boole – and insisting on showing every possible intersection. Venn diagrams are really just a special case of Euler diagrams.*

This brief historical sojourn highlights the sheer versatility of Venn and Venn-like diagrams. Our aims may not be as lofty as Llull's desire to understand the workings of God, but logic diagrams can bring visual clarity to how a situation is viewed by different people.

---

* Venn himself referred to them as Eulerian circles.

Once we identify the logic diagram in play, we can step back and ask if there is a more suitable alternative.

Take the idea of organisational structure, and the question of how best to manage a workforce with varied skill sets. A traditional view is to neatly package workers into departments based on their skill – the marketing team over here, sales over there, the tech workers in some far-flung corner of the building where they won't be disturbed. We can think of this structure as a particular type of Euler diagram, with no intersection between departments:

Sales	Customer Support
	HR
Marketing	Technology

This structure has a name: MECE, which stands for *mutually exclusive and collectively exhaustive*. In other words, no team member belongs to more than one department, but everyone slots in somewhere. MECE was coined by Barbara Minto of the consultancy firm McKinsey in the 1970s.[7] As well as being the first woman to earn an MBA, she embedded MECE as a general problem-solving framework for management consultants. The idea is to break a problem into a collection of MECE hypotheses, each of which can be addressed separately. You can apply the same approach to presentations by splitting a slide deck into a set of MECE sections, or you might offer your customers a range of MECE services.

Mathematicians have another name for MECE: *partitioning*. Through the lens of our various logic diagrams, we can see that MECE is one approach among many, and a decidedly reductionist one at that. MECE is appropriate when problems are truly separable – for instance, if you want to understand why your website is losing customers, you might set up a funnel that tracks each stage of activity, to see where the biggest drop-off occurs. You can then focus your design resources on strengthening that aspect of the user journey.

But not all problems are so easily reducible to their constituent parts. The human brain cannot be understood by analysing individual neurons, while the intelligent behaviour of ant colonies transcends the know-how of any individual ant. A complex problem, similarly, may require the synthesis of multiple perspectives. One of the most famous problems in maths is Fermat's Last Theorem, which says that the equation

$$x^n + y^n = z^n$$

has no positive whole number solutions when n is greater than 2. This claim was first made in 1637 by the French amateur mathematician Pierre de Fermat, who added that he had a proof but was not able to fit it into the margin of his notes. It would take mathematicians another 358 years to finally prove the theorem (Fermat's proof, if it even existed, would be markedly different and, most likely, flawed). When the mathematician Andrew Wiles finally discovered a proof, he used a specialised object known as modular elliptic curves, which brings together concepts in fields such as algebraic geometry and number theory. It was a mathematical mash-up for the ages, spawning new areas of research as these two unconnected fields danced in lockstep for the first time. The complex problems shaping the contours of scientific research tear away artificial barriers between disciplines. Mainstream education, on the other hand, is a giant nod to MECE, with only lip service paid to the crossovers between subject areas – curriculums would do well to infuse this interdisciplinary spirit.

We might say a MECE organisational chart is the *least Venn-like* of schemes; a more Venn-like one embraces overlap between departments. One alternative to MECE structures is the so-called 'team of teams' model,[8] in which every major project a company undertakes has at least one representative from each department, engendering a shared sense of purpose as ideas penetrate departmental boundaries.

Team of teams has proven effective in modern warfare, which is becoming ever more complex and unpredictable due to the speed and agility with which adversaries are able to plan and communicate. Top-down military hierarchies are no longer fit for purpose; lower-level units instead need the autonomy to make quick decisions and to adapt to rapidly changing conditions on the battlefield in a manner that the rigid feedback loops of MECE prohibit.

Team of teams is also a signature quality of innovative companies, where the top-down rigidity of MECE gives way to a culture of knowledge-sharing. To give a concrete example, many organisations purposefully design their office space to bring about spontaneous 'collisions' among colleagues (a practice made famous by the R&D company Bell Labs). Collaboration is more organic, while intelligent and creative behaviour emerges as more than the sum of workers' individual parts.

Not everyone takes well to the concept of overlapping structures. Venn diagrams may be the only mental model in this book that was literally banned by a government. After seizing power by a coup in Argentina in 1976, the ruling military junta forbade schools and textbooks from including Venn diagrams. The fascist regime did not take well to overlapping circles that could be interpreted as a model of social collaboration, threatening its grip on power. The Venn diagram, it feared, might become a symbol of organisation or resistance.

By now you should be persuaded that there is nothing to fear, and that ideas of intersectionality are not as contentious as culture war warriors, or military dictators, would have you believe. The Venn diagram is the most benign of tools that encourages thinking beyond singular reference points and shows the power of combining our perspectives with other, overlapping worldviews.

5

Axioms

How core beliefs underpin our worldview

In 2022, the introduction of 'low-traffic neighbourhoods' in Oxford drew me into the most uncomfortable of alliances. Through the erection of bollards, motorised vehicles are prevented from taking shortcuts through densely populated residential areas. As a frequent visitor to the city, I've joined the growing chorus of disapproval of the policy, on the basis that it prolongs bus journeys and displaces traffic to nearby areas. Large numbers of social media users seemed to agree with me, but my enthusiastic nods towards posts denouncing the policy came to an abrupt halt when I learned that they were being penned by far-right extremists.

Opposition to Oxford's LTNs stretches far beyond the City of Dreaming Spires and has morphed into a conspiracy theory concerning a bid to curtail citizens' movements.[1] This theory conflates LTNs with the concept of 'fifteen-minute cities', a policy with the benign aim of allowing residents to access crucial services and amenities within a fifteen-minute walk. And at the root of this particular protest movement is a fundamental distrust in government.

LTNs are a perfect illustration of how the common ground we find with others can stem from fundamentally different beliefs; my own opposition had nothing to do with my level of trust in government, and everything to do with practical concerns around residents' daily commute. The notion that a government's sole purpose is to maintain control over its citizenry, mixed in with other paranoid beliefs such as the existence of a cabal of global elites intent on forming a new world order, may occasionally intersect with more conventional views on urban planning policies, but they also feed the conspiratorial frenzy around vaccines, demographic shifts and the supposed rise of global governance. The 'Like' button should be accompanied by a warning to check the source of the content you're approving – it may derive from an ideology that you may otherwise deem problematic.

This chapter makes a distinction between our derived beliefs and the core beliefs from which they stem. In mathematics, core beliefs are called axioms. Mathematicians pride themselves on the grunt work that comes with proving their claims, and the rigour with which they lay out their arguments. But even mathematicians have to rely on a handful of axioms to get their subject off the ground – statements so basic and self-evident that they do not require any proof.

Axioms are the underpinning of every mathematical system. Once they are cast, logic will take care of the rest – through logical inference alone, those axioms will combine into new results. Those new results combine further, cascading into an entire body of knowledge.

The first attempt to define a system in such terms was made by the Greek mathematician Euclid; some 2,000 years ago, he wrote down five axioms that give a comprehensive theory of geometry. One of them tells us that we can connect any two points with a straight line, another that all right-angles are equal, another that parallel lines exist. The axioms for the number system were spelled out much later, specifying such things as 'The order in which you add two numbers does not change the outcome' or 'There is a number, called zero, that has no effect when added to any other number'. The chronology here is important; mathematicians often work back from an established body of knowledge to define what they think is the most appropriate collection of axioms.

This chapter will review ill-fated efforts to establish a perfectly logical mathematical system by applying a handful of axioms to sets (which, as you'll recall, are just collections of objects). But it is not just a historical reckoning of our failed attempts to achieve abstract mathematical supremacy; being aware of our axiomatic real-world beliefs is essential not only to help us avoid conspiracy theories, but also to identify the root cause of everyday disagreements. For instance, debates about the role of government, from taxation policies to regulation, can be viewed in terms of core beliefs concerning *responsibility*. For a libertarian who places responsibility squarely in the hands of the individual, it stands to reason that government intervention should be kept to a minimum. Those who advocate a more interventionist role for government instead view responsibility as belonging to the collective, from which greater investment in social welfare is an obvious consequence. The differences that manifest on a policy-by-policy basis can give an exaggerated impression of our political divide when, at root, these differences are natural by-products of a handful of cherished beliefs.

The ability to trace our view of the world to a handful of governing beliefs can be incredibly liberating once we recognise that axioms are mutable – and that engaging with different axiomatic beliefs opens our minds to alternative viewpoints. That may seem a strange reflection for a mathematician to make, but even mathematics cannot be reduced to a single collection of axioms – the perfectly logical system, it turns out, is unattainable. But what, exactly, does it mean for a system to be *perfectly logical*?

Consistency and completeness

Mathematicians carry two hopes for their axiomatic systems. The first is *consistency*, which means that the axioms cannot be combined in such a way that leads you to conclude that a statement is both true and false. A contradictory statement is a telltale sign of a theory gone wrong. Consider two of the conspiracy theories that circulated during the Covid-19 pandemic: one that said the risk posed by the virus was overblown ('It's just a common cold!' or even 'The pandemic is a hoax!'), and the other claiming that the virus was deliberately engineered as a bioweapon by China to inflict harm on the West. You can subscribe to one belief or the other, but to accept both at once is nonsensical – 'post-truth' taken to its absurd extreme.[2]

Once you allow a single contradiction into your system, it acts as a nerve centre for misinformation, whereby every other statement can be shown to be both true and false at the same time. This is known as the 'principle of explosion'; one contradiction opens the floodgates to others, rendering your system unable to distinguish truth from falsehood.

A mathematician therefore seeks to avoid contradictions at all costs, yet building a system without them is not straightforward. It takes little effort to craft absurdities with plain English. Consider the statement: *this sentence is false*. If the statement is true, then, by its very meaning, the sentence must be false, which makes it true. It's a head-spinner, but the break in logic is clear: we have a statement that, when assumed to be true, implies its own falsehood, and vice versa. The philosopher Karl Popper's 'paradox of intolerance' is a tangible example of a real-world concept that turns on itself in this way. Popper argued that a perfectly tolerant society would have to give space to intolerant people who, left unchecked, will exercise their freedom to attack and destroy tolerance. Popper thus concludes: 'In order to maintain a tolerant society, the society must be intolerant of intolerance.'[3]

This clever wordplay posed a challenge for mathematicians who believed that sets could secure the most rigorous mathematical foundation of all. Bertrand Russell was able to formalise these paradoxes in terms of sets. He considered a set, S, whose elements are *all the sets that do not contain themselves*, before asking: *does S contain itself?* After a bit more head-spinning, we end up in the same situation: if S contains itself, then it doesn't, but if it doesn't, it does.

The remedy was a handful of agreed-upon set theory axioms, some of which we glimpsed in the previous chapter in our exploration of Venn diagrams (such as the axiom of union, which says that we can combine the elements of two sets to form a larger set, and the axiom of intersection, which says we can define yet another set by taking the elements common to both sets). The set theory axioms were defined in such a way that these types of paradoxes were avoided. In essence, the axioms imposed a restriction on what counts as a set, disqualifying anything as ungainly as Russell's example. The axioms appeared to set mathematicians on the path to a perfectly consistent system.

The second property mathematicians desire of any system is *completeness*, which means that any statement expressible in the system can be proven true or false from the axioms. For a system to be *incomplete*, it would have to entertain so-called 'undecidable' statements – those that, as far as the system can demonstrate, are neither true nor false. It would be akin to a theory of geography that can't tell you whether Paris is the capital of France, or a theory of history that can't tell you whether the Allies triumphed in the Second World War. Mathematics, a subject renowned for its decisiveness, ought to be free of any such unknowns.

Some systems, such as Euclid's geometry, satisfy both requirements; his axioms give rise to a theory that is both contradiction-free and accounts for the truth or falsehood of every possible statement regarding the geometry of the plane. The ambition of set theorists was to achieve the same outcome for a more comprehensive system – an ambition that, as we shall see, was doomed all along.

The axiom of choice

Not every axiom is greeted with unanimous approval. The opprobrium generated by Cantor's multiple infinities, described in the previous chapter, had much to do with his contemporaries' refusal to accept a particular axiom – the aptly named *axiom of infinity*. In brief, this axiom allows for the existence of a set with infinitely many objects. The idea rubbed against the usual way mathematicians thought about logic, which was rooted in finite collections of objects. Cantor's ideas would eventually prevail, as the mathematician David Hilbert foretold when he proclaimed: 'No one will drive us from the paradise which Cantor created for us.' Cantor's counterintuitive infinities were welcomed by mathematicians because they blended into a fully coherent framework for how to use sets to build mathematics from the ground up.

The debate on infinity may be settled, but if you wish to stir up trouble among a group of mathematicians, ask them if they accept the *axiom of choice*. You may not witness an all-out brawl, but you might notice some uncomfortable shuffling as they contemplate a long-standing mathematical controversy.

At first, the axiom of choice seems as obvious as the other foundational assumptions that mathematicians subscribe to. To understand it, imagine you have ten jars in front of you, each packed with sweets. You are asked to draw a single sweet from each jar, which poses no dilemma – just dive into each jar in turn, using whatever selection process suits you.

The axiom of choice says that you could undertake your selection regardless of the number of jars – even if there were *infinitely* many. In some hypothetical situations this would remain a straightforward task. If you know, for instance, that each jar contained a single red sweet and you opt for that one each time, your selection mechanism will scale easily to the infinite case. But if there is no obvious candidate in each jar, could you define a method for selecting a sweet? Perhaps you would opt for the largest – but what if the sweets in a particular jar are all the same size? Or you might choose the highest placed sweet, but in some jars multiple sweets may jostle for this position, making the selection ambiguous.

To invoke the axiom of choice is to presume that some selection mechanism always exists, regardless of the number of sets you are choosing from. Tantalisingly, the axiom does not give any insight into how to make your selections – only that it is possible to do so. For another everyday example (borrowed from Bertrand Russell), you could easily describe your rule for choosing a shoe from an infinite number of pairs – it could be: 'Choose the left shoe every time.' But for infinitely many pairs of matching socks, there is no obvious rule you could apply in advance. The axiom of choice is your abstract workaround; it tells you the selections are still possible, without specifying how.

As trivial as this axiom may seem, many mathematicians depend on it. The chapter on dimensionality in this book would be lost without it; mathematicians have shown that the axiom of choice is logically equivalent to the assumption that every vector space has a specified dimension (see Chapter 8). Advocates of the axiom argue that it's a natural extension of the finite case; if you are untroubled choosing from ten jars or one million, then surely you can accept that there must be some selection mechanism for infinitely many. Isn't this, after all, the same kind of game we play in calculus, breaking intervals into infinitesimally small chunks and assuming the limit exists?

However, one mathematician's 'workaround' is another mathematician's 'cheat code'. The axiom of choice draws two main objections, of which the first is rooted in its non-constructive nature. In one view, mathematical objects can only exist if there is an explicit description of how to create them, which the axiom of choice conveniently (or inconveniently, depending on your view) omits.

The second objection has to do with some questionable – not to mention downright bizarre – consequences that the axiom gives rise to. It also happens to be equivalent to saying that every set can be put in some definite order, taking our instinct to order things up to its extreme. Just consider the set of all real numbers that occupy the number line; our usual way of ordering them doesn't do the job because, for one thing, there is no smallest number – you can always find one to the left of a given candidate. The axiom of choice says 'Don't worry, there's another way to order them, but I won't tell you what it is.' You might have sympathy for those reluctant to take such claims on blind faith.

As for the bizarre, the axiom of choice logically leads to the Banach–Tarski paradox, which states that you can turn one ball into two exact copies of itself. The more technical description is that it is possible to split a three-dimensional ball into several pieces that do not overlap and then reassemble the pieces without any stretching or bending in such a way that you end up with two balls each identical to the first. This result is pure wizardry, yet if one assumes the axiom of choice, then the second copy of the ball can be magicked into existence.

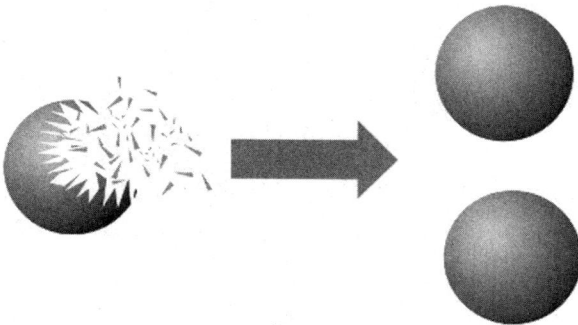

There is a catch, as you might expect: in practical terms, you'll never actually be able to assemble the two balls – the 'pieces' are so irregular that they fail to resemble anything we could carve, and their 'volume' changes depending on how the assembling occurs. The result is the epitome of abstraction.

You might think there is a logical way to resolve the debate. This is mathematics, after all – surely one way or another we ought to be able to establish whether or not the axiom of choice follows logically from the rest of the set theory axioms that were already agreed upon. However, two discoveries show that the debate is impossible to resolve. On the one hand, the logician Kurt Gödel showed in 1938 that, if the other set theory axioms are consistent, adding the axiom of choice preserves this consistency. This means, for one thing, that the axiom of choice cannot be proven false from the other axioms. But in 1963 the mathematician Paul Cohen showed that, if we assume the other axioms to be consistent, the axiom of choice cannot be proven true from them, either.

Taken together, these two results show that the axiom of choice is independent of the others. We can add it to our system or ignore it – either option is logical. So we're in the undesired territory of having an incomplete system, with a statement that is undecidable. Mathematicians have grudgingly accepted the subjectivity implied by the axiom of choice. Many who invoke it do so with lingering unease.

Perhaps one reason the axiom of choice rankles so much is that it impinges on our understanding of human behaviour, daring to presume that we can make choices without thinking about *how* they are made. This brings it into proximity with the field of behavioural economics, which is predicated on a raft of axiomatic beliefs concerning how we make decisions. The theories often seem plausible at first, only to be contradicted in later experiments.

Let's turn to the idea of humans as *rational agents*. The classical view in behavioural economics is that we make decisions to maximise what theorists term 'utility' – how beneficial a certain outcome is.[4] When we express a preference for pizza over curry, it is because a 'utility function' assigns a higher value to the former. There is rigorous mathematical proof to back this up, yet something is awry, because humans are not always tuned to sound choices. We are afflicted with cognitive biases, leading to bad decisions that point to a surrender of our rationality.

You will find no fault in the logic of economists' mathematical arguments. That only leaves us with the axioms, which we should always place under the microscope when the conclusions of a theory are at odds with the evidence. The axioms driving the instrumental view of human behaviour include, for example, the *axiom of completeness*, which says you will always have a preference between two outcomes (or judge them to be equally good), and the *axiom of transitivity*, which says that preferences follow a transitive logic (meaning that if you prefer cats to dogs and dogs to rabbits, then you prefer cats to rabbits). In Chapter 3 we saw examples where humans do not neatly order things; behavioural economics has stumbled not because of flawed logic but because of its misplaced founding assumptions. One is entitled to cling to these axioms but must then accept that they are describing a very different reality to the one originally envisaged, a kind of fairy land with alien creatures that think and behave in a predictably blunt manner.

Our reluctance to abandon axioms leaves us with a trade-off between elegant theories and their unsettling logical outcomes. That is the essence of axioms; we have to own them to their logical end, however unpalatable. Mathematicians who embrace the axiom of choice also have to accept the strangeness of the Banach–Tarski paradox. If we deem the consequences unpalatable, we have no option but to rethink the assumptions on which they are based. Mathematicians only ever prove things with respect to their underlying axioms. Most are accepted without contest, but they occasionally ruffle feathers.

Every axiom is an assumption of some kind and if we accept them without scrutinty we may find ourselves hopelessly constrained. A mathematical example: suppose I decide I've had quite enough of the standard framing of numbers and insist on adding another to the mix:

Mubeen Axiom:
Multiplying any two numbers gives zero.

At one stroke, this would tear to pieces what we take for granted about numbers. Possibilities abound and theorems will be proven. Perhaps one day we'll find practical use for this most abstract of number systems, in much the same way that modular arithmetic (see Chapter 3) has found applications in timekeeping and cryptography.

There is a slight hitch, however. When I start to play with my axioms, combining them with a few of the standard ones, I soon realise that every number in my system must itself be zero. That's because, starting with any number, which I'll call Y,

$$Y = 2 \times 0.5Y$$

and since this is the product of two numbers, I am forced to conclude that Y is zero. Whatever results I might go on to prove in my system are redundant because it is barely a system at all – it is what mathematicians would call a *trivial* system, one that is lacking in any meaningful content or structure (in my case, the system contains only the number zero). My undergraduate tutor once regaled me with the tale of a PhD student who spent months proving theorem upon theorem concerning a geometric system he had devised, only to realise subsequently that his axioms amounted to a similarly trivial system. The theorems remained true, but as they only applied to the simplest of structures they were stripped of significance.

Axioms should have a generative effect, producing a swathe of new results that combine and give rise to rich theories. The reason trivial systems never develop into anything more meaningful is that their axioms have a collective destructive effect, existing in such tension that they can only be applied to the most uninteresting of cases. Trivial systems represent a gross oversimplification of complex realities, as exemplified by the 'rational agents' of behavioural economics that portray humans as relentless utility optimisers.

Trivial systems also help us understand how conspiracy theories persist, despite their apparent contradictions. The beliefs driving them are often vast in scope. Belief in a malevolent global government, for instance, becomes a blunt instrument through which to analyse migration policies, attempts to curb greenhouse gas emissions, or methods to prevent the spread of a deadly pathogen. Such policies are readily interpreted as attempts at population control. The belief drives a paranoid worldview where nobody or nothing can be trusted, which makes any dissenting evidence easy to dismiss. In the mind of a conspiracy theorist, therefore, everything feels coherent and there's no contradiction to be found. Yet a conspiracy theory has almost no explanation of how the world actually works – it is unmoored from evidence and unaccommodating of scientific inquiry. There may – if we're being charitable – be some internal consistency to a conspiracy theorist's ideas, but they amount to the most trivial and banal of belief systems.

Well-crafted axioms protect against the threat of contradictions leaking into a system, but they can also prove overly restrictive. As an amateur card magician, I dislike tricks that are based on 'gaffe decks' – sets of cards that have been manipulated in some significant way; they do too much heavy lifting for the magician and lack a certain elegance. I also lack enthusiasm for tricks involving counting or spelling (they feel too contrived), and for tricks that rely on deft sleight of hand (they privilege dexterity over audience interaction). These choices are arbitrary; personal preferences that lend character to my card tricks. Adhering to them all is a challenge; it requires serious thought to devise a trick that meets every requirement while also producing a genuine *wow* moment. My tricks usually reflect a compromise: some sleight of hand here, a prearranged deck there. Of course, professional magicians, like all artists at the top of their field, conjure up their brilliance under the harshest of constraints. But amateurs like me are forgiven for relaxing the constraints, and in doing so are granted a wider arsenal of tricks to take to our unsuspecting audience.

Incompleteness

Let's return to the ultimate prize mathematicians were seeking: a collection of axioms that give rise to a system that is both free of contradiction (consistent) and that can be combined to demonstrate every expressible truth in the system (complete). Mathematicians thought they were onto a winner with their agreed upon axioms for sets, but then came the discovery of the axiom of choice, and its unsettling consequences. One option is to take the axiom of choice on the chin: embrace it as one of your ground truths and reluctantly accept consequences like the Banach–Tarski paradox. This new system may be bizarre in places but that's a price worth paying if it delivers on the promise of consistency and completeness.

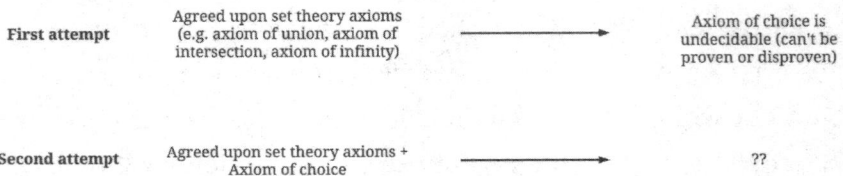

| First attempt | Agreed upon set theory axioms (e.g. axiom of union, axiom of intersection, axiom of infinity) | ⟶ | Axiom of choice is undecidable (can't be proven or disproven) |

| Second attempt | Agreed upon set theory axioms + Axiom of choice | ⟶ | ?? |

Onwards you go, using the axiom of choice, along with those already agreed set theory axioms, to dispatch one proof after another. Unfortunately, another problem arises along the way, and it comes from Cantor's idea of multiple infinities. The first two levels of infinity, you will recall, are the *countable* infinity of whole numbers and the *uncountable* infinity of 'real numbers' that populate the entire number line. Inquisitive as you are, you might wonder if there is another level in between. Using the concept of cardinality (from the previous chapter), we can frame this in more precise terms. Were such a set to exist, two things would be true:

- Its cardinality would be larger than the cardinality of the set of whole numbers. That is, we can find a way to pair every whole number with a different element of this set, but every pairing mechanism misses out some elements of our candidate set.
- Its cardinality would be smaller than the cardinality of the real numbers. That is, we can find a way to pair every element of our set with the real numbers, but every pairing mechanism misses out some real numbers.

Does this in-betweener set exist? The answer doesn't leap out – there's no set that obviously fits the description. But you have come armed with your axioms – including the axiom of choice – and figure that the problem will eventually be solved one way or another (if not by you, then by a seasoned mathematician). It is the most binary of questions: either such a set exists, or it manifestly does not. Your axioms should guide you to whichever outcome is true.

Don't hold your breath. Mathematicians have shown that no such set will ever be found from your axioms. But that's not all: assuming the set does exist will not in any way contradict your axioms, either. We are in another logical cul-de-sac, facing a statement that cannot be proven or disproven from our set theory axioms (which now include the axiom of choice). Just as the axiom of choice was found to be independent of the standard set theory axioms, this new result – the so-called *continuum hypothesis* – is independent of the updated system that tagged on the axiom of choice. Even invoking this axiom is insufficient to establish the truth or falsehood of the continuum hypothesis. We're back to the quagmire of an incomplete system.

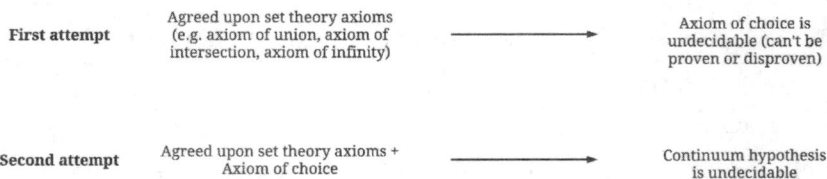

First attempt	Agreed upon set theory axioms (e.g. axiom of union, axiom of intersection, axiom of infinity)	⟶	Axiom of choice is undecidable (can't be proven or disproven)
Second attempt	Agreed upon set theory axioms + Axiom of choice	⟶	Continuum hypothesis is undecidable

The continuum hypothesis was announced as one of David Hilbert's twenty-three great unsolved problems at the turn of the twentieth century. Embedded in it was the hope that mathematicians would eventually arrive at a complete and consistent system – one that would determine, among other things, whether those in-between infinities exist. If there was such a version of mathematics, it would have to derive from a new set of axioms that didn't give rise to these 'undecidable' statements.

The tragic revelation for mathematicians was that there is no such panacea. We cannot blame the standard set theory axioms, or even the axiom of choice, for the occurrence of undecidable statements. It was Kurt Gödel who demonstrated that *any* consistent system basic enough to contain the rules of arithmetic will always contain statements that can be neither proven nor disproven.

This is the first of Gödel's *incompleteness theorems*, and it forces mathematicians to choose between consistency and completeness. If you cling on to the logical certitude of mathematics, then you must surrender any ideal of a grand unified theory; some statements are guaranteed to remain undecidable within your system, neither true nor false.

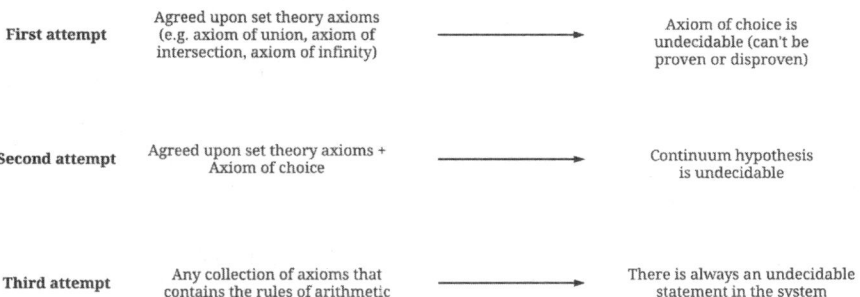

First attempt	Agreed upon set theory axioms (e.g. axiom of union, axiom of intersection, axiom of infinity)	———————▶	Axiom of choice is undecidable (can't be proven or disproven)
Second attempt	Agreed upon set theory axioms + Axiom of choice	———————▶	Continuum hypothesis is undecidable
Third attempt	Any collection of axioms that contains the rules of arithmetic	———————▶	There is always an undecidable statement in the system

Gödel's theorem takes us to the edge of what we can know. It leads us to the realisation that axiomatic systems, however engineered, are unable to span all truths, except in the most basic of systems. If that is true of the abstractions of set theory, it is certainly true of the axiomatic beliefs that shape how we see the world.

This tension between consistency and completeness is often at play in the real world. It is exemplified by the obsession that many industries have with measurement.[5] Adopting the dictum 'If you cannot measure it, you cannot improve it', many companies resort to quantifying performance in terms of numbers. The paradigm is also prevalent in education, where test scores are used to judge everything from students' potential to the performance of teachers, schools and entire education systems.

Metrics are appealing because they appear objective and without contradiction. Yet our instruments of measurement – a company's key performance indicators or a student's exam grades – only reveal so much about our potential. For everything that metrics do capture, they gloss over much more. A more progressive approach to evaluation is one that takes into account a range of other factors, qualitative as well as quantitative. The obvious critique of this more holistic approach is that it is too subjective; without an empirical basis, it is not beholden to the same scrutiny as metrics. That may be true – proponents of this approach would have to concede that it contains unfillable gaps – but if Gödel's incompleteness theorem teaches us anything, it is that the grand unified theories are unattainable anyway. Incompleteness is a feature of all but the most prosaic systems.

Gödel delivered another blow to the idea of a perfect logical system with a second incompleteness theorem that shows it is impossible for a system to demonstrate its own consistency (unless, once again, it is so basic that it doesn't contain the rules of arithmetic). The mathematician André Weil may have put it best when he said: 'We know that God exists because mathematics is consistent, and we know that the devil exists because we cannot prove the consistency.' So even the one thing that paradigms such as measurement might have going for them – that they are consistent – is disputable. Perhaps the only reason metric-based evaluation systems appear free of contradiction is that they are void of any real structure.

Gödel's second incompleteness theorem resigns us to some degree of blind faith. If we sought comfort in the idea that our beliefs are consistent, we would have to accept that this is not something we could establish for certain, except for the most basic of beliefs. It's why you will never be able to demonstrate, for instance, that you have the perfect model for how to run a society (recall from Chapter 3 that if your model is to include a perfectly fair voting system, then you are already reduced to a dictatorship).

The diversity of human thought owes much to the fact that our core beliefs, so rich and varied, can be difficult to falsify. Questions such as where responsibility resides or how to evaluate progress are not always amenable to scientific interrogation. We work hard to construct belief systems that are internally consistent, but dogmatic attachment to any one set of beliefs can blind us to valid alternatives.

Mathematicians learned this lesson by realising that Euclid's axioms for geometry are not the only way to think about shape and space. For instance, we can conceive of geometries that do away with the axiom about parallel lines; in the geometry of a sphere, lines correspond to all possible equators (circles that split the sphere into two equal parts). In this new geometry, there are no parallel lines. Euclid's geometry is perfectly consistent, but his axioms apply only to flat planes and not to curved surfaces. It works in most everyday situations; when arranging the furniture in my office, I'm content to invoke the idea of parallelness as I line up my desk with my bookshelf. When I'm on a long-range flight, however, I am reliant on the geometry of the sphere to find the most efficient route to my destination.

In mathematics, axioms are made to be tinkered with – a creative act that abandons conventional wisdom as we contemplate unforeseen ways of thinking. In the real world, it also amounts to an act of empathy; by entertaining counterfactuals, we can learn that multiple, competing viewpoints can all be valid at once.

If the set theorists had achieved their aims, the axioms would be set in stone, with no room for debate, dispute or compromise. That mathematics, a subject as poised as any to live up to this promise, came up irredeemably short, is proof of the subject's creative spirit.

6

Logic

How to win (and lose) arguments

My strongest pitch for mathematics is that it helps us to argue more effectively. In fact, I'm tempted to promise that this chapter will help you win every debate, but I know you are not so naïve as to think humans defer solely to cold logic. Even so, if mathematics is conditioning for the mind, logic is the high-intensity training that pushes us to notice the details of every claim and counterclaim.

The last chapter introduced us to statements whose truths we treat as self-evident, but we also need mechanisms for combining axioms and establishing the truth or falsehood of more sophisticated statements. This is where logic comes in; it's the fuel that sets mathematics in motion. It grants us legitimacy when we move from one proposition to the next, and to use what we already know to arrive at new insights.

The mechanism we will focus on is called *modus ponens* (Latin for 'method of affirming') and it offers a clear plan for how we can demonstrate that a proposition is true. Suppose we manage to find a premise (which we'll call P) and a desired conclusion (which we'll call Q), and that we have some way of showing the following:

(a) P is true, and
(b) If P is true, Q is true (we'll use the shorthand P ⇒ Q, where ⇒ should be read as 'implies'.)

According to *modus ponens*, we can infer that Q is true, which was our aim all along. A simple example:

(a) It is Friday
(b) On Fridays, I spend the morning in a coffee shop (It is Friday ⇒ I will spend the morning in a coffee shop)

We can conclude that I am about to spend the morning in a coffee shop. In this example, the conditions (a) and (b) are easy to establish, provided you know the day of the week and what you get up to on Friday mornings. We don't always have this luxury. We may not even be clear about what premise to choose, and once we've identified a plausible candidate, work remains to show that, firstly, P is true, and, secondly, that P ⇒ Q.

Showing that P is true may itself require another application of *modus ponens*, where P now becomes the conclusion and we are in search of another premise. For example, imagine you are driving down the motorway when it suddenly occurs to you that you may have left your wallet at home. Keen to avoid any disruption to your journey, you desperately try to convince yourself it is in your possession (which is your statement Q). You might reason as follows:

P	The jacket I placed on the passenger seat contains my wallet
$P \Rightarrow Q$	If my jacket contains my wallet, then the wallet is in my possession

By *modus ponens* you can rest easy because the desired conclusion Q now follows. But a few miles further along your journey, you are struck by another thought: *what if I didn't place my wallet in my jacket?* Your mind now scrambles to retrace your steps and determine if you did, in fact, remember to place your wallet in your jacket. This statement, which moments earlier you took as your premise, is now the conclusion (let's call it Q') of a second argument:

P'	My jacket pocket is bulging
$P' \Rightarrow Q'$	If my jacket pocket is bulging, then my wallet must be in the jacket

A quick glance over at the passenger seat confirms that the pocket is indeed sticking out, so you can now be sure it contains the wallet (this assumes, of course, that you have not placed another hefty item in the pocket that you are mistaking for the wallet). Your earlier conclusion applies after all – you have the wallet with you. It just needed a second argument to confirm the premise.

Implication

Mathematicians have crafted clever ways of proving their claims, which typically comprise multiple iterations of *modus ponens*. We'll look at two methods of proof – proof by contradiction and proof by induction – later in this chapter, but let's first explore the second condition of *modus ponens*, the *implication*, which packs in more meaning than might first seem apparent.

In everyday use, the word 'imply' carries some subjectivity. 'He's a hard-working student, which implies he'll do well in his exams' is a plausible sentiment that will hold up much of the time, but there will be exceptions. The student might have an off day or study the wrong material, resulting in a below-par performance. In the mathematical sense of the word, an implication is free of such ambiguity. The claim

$$x \text{ is greater than } 1 \Rightarrow x^2 \text{ is greater than } 1$$

holds up without exception for numbers: there is no number greater than 1 whose square is not also greater than 1. On the other hand, the statement

$$x \text{ is a positive number} \Rightarrow x \text{ is greater than } 5$$

fails because, while most positive numbers are greater than 5, a few are not. It takes a single point of failure to render the implication invalid.

In real-world formulations, context makes all the difference. My claim that

$$\text{I am in my personal office} \Rightarrow \text{I am at home}$$

is a valid implication if my only office – of any description – is in my home. But if I count coffee shops as another type of 'personal office' (not implausible, given how much time I spend in them), the implication no longer holds up; there are many instances where I can be found in a coffee shop while not at home.

A common mistake, known as the 'post hoc fallacy', is to assume $P \Rightarrow Q$ just because Q occurs after P. If I eat ice cream and end up with a stomach ache, this doesn't mean the ice cream caused the pain. There are many other possible causes – if I got punched in the stomach soon after my ice cream, then I already have at least one other potential explanation (along with countless causes that may remain hidden). It is also true that P and Q can occur in reverse chronological order, such as in medical diagnosis:

He has a high fever $\Rightarrow$ He caught a bug

This may be a valid implication, yet clearly the logic works backwards through time: first the observation of the fever, then the investigation of the patient's recent history to track down the cause.

Some more terminology: whenever $P \Rightarrow Q$, P is said to be a *sufficient* condition for Q or, equivalently, Q is a *necessary* condition for P. They are two ways of saying the same thing – that every time P is true, Q must also be true. It is, of course, possible for P to be both sufficient and necessary for Q, in which case Q will also be sufficient and necessary for P; the two statements are equivalent. The shorthand is:

$$P \Leftrightarrow Q$$

My earlier statement

I am in my personal office $\Rightarrow$ I am at home

is not an equivalence: I can be at home (in the kitchen, say) but not in my personal office. The following equivalence, on the other hand, holds true because bananas are a staple part of my breakfast, but I will not consume them under any other circumstances.

I am eating a banana $\Leftrightarrow$ I am having breakfast

These distinctions, as minor as they are, can catch people off guard. Suppose your friend tells you they will reward you with dinner if you babysit for them that evening. You agree to the trade, which we can formulate as:

You babysit for your friend $\Rightarrow$ They buy you dinner

If their plans change and they no longer require your babysitting services, are you still entitled to a free dinner? The logical answer is *no*; you have not been given the opportunity to meet the sufficient condition for earning your dinner. Then again, since the agreement says nothing of what happens if you *don't* babysit, you may take your chances and ask for dinner anyway. If your friend had said, 'I'll buy you dinner if and *only* if you babysit for us', then you would have all the clarity you need: no babysitting means no dinner. There would be an equivalence:

You babysit for your friend $\Leftrightarrow$ They buy you dinner

The precision demanded by logic can make it feel abstract and of little everyday value. But on the contrary, much of logic's power comes from the ability to frame statements in different ways. Consider, for example, the implication:

You generally feel lethargic $\Rightarrow$ You are
not exercising enough

Let's accept this diagnosis as valid (though there may
of course be other explanations). It's not a particularly
uplifting message – it may even come across as judge-
mental – but consider the alternative formulation:

Exercise regularly $\Rightarrow$ Your energy levels will rise

This framing evokes a different set of emotions; it touts
the benefits of exercise and promises a healthy reward.
Yet in logical terms, the statement imbues the same
meaning as the previous version. The second statement is
known as the *contrapositive* of the first. Every statement
of the form $P \Rightarrow Q$ has a contrapositive that preserves its
meaning, which takes the form:

Q is false $\Rightarrow$ P is false

So every implication can be presented in two ways, and
our choice can make a material difference to our outlook. Which of the following two statements seems more
hopeful?

- If I fail this job interview, it will be because I am
  unprepared.

- If I prepare adequately, I will succeed in this interview.

The first statement dwells on potential failure; the second encourages a more proactive mindset and pushes for action. The two statements are logically equivalent, but they land differently.

Vacuous statements

The idea of implication is to get from one true statement to another. Our conclusion, Q, depends on the premise P being satisfied. But this invites the question: what if P can never be true? There are four scenarios in play, depending on whether P and Q are each true or false, which we can summarise in a 'truth table':

P	Q	$P \Rightarrow Q$
True	True	True
True	False	False
False	True	True
False	False	True

It's a slightly strange notion, but if P is false, the implication $P \Rightarrow Q$ is valid irrespective of whether Q is true. This means, for instance, that the following implications are valid:

Paddington Bear is the prime minister of
the UK $\Rightarrow$ London is the capital of England

Paddington Bear is the prime minister of
the UK $\Rightarrow$ London is the capital of Greece

A false premise implies anything and everything, true and false. There's only one way that $P \Rightarrow Q$ can fail: when P is true and Q is false. Here's a fun test of whether you grasp the concept. Suppose you are presented with four cards that display the characters A, D, 4 and 7, and are told that every card has a number on one side and a letter on the reverse. You are also told that if a card has a vowel on one side, it should have an even number on the other side. Which cards do you need to flip over to determine which cards might violate this rule? Take a moment to think this through.

Studies have shown that the card with A is the most popular choice, followed by the one with 4.[1] There is no doubt about the former, but the card displaying 4 should be ignored because it cannot fail: if its opposite side has a vowel, the rule works, and if the opposite side has a consonant, the rule has not been violated because it says nothing at all of consonants. The other card you need to flip is the 7; if its opposite is a vowel, the rule has indeed been violated. D can safely be ignored – it isn't a vowel and so poses no risk of breaking the rule.

Studies like this explain what has become known as 'confirmation bias', the tendency of humans to seek out evidence that supports a belief ahead of evidence that refutes it. We struggle to intuit the idea that an implication is automatically true when the premise is false.

When $P \Rightarrow Q$ is true because P is false, the implication is said to be 'vacuously true' – it holds only because it has no chance of failing. My claim to have completed every Iron Man competition I've ever entered in under twelve hours may sound impressive and I can assure you it's true – but only because I've never attempted an Iron Man.

Vacuous statements sometimes take the guise of unfulfilled promises. A boss in an old job once pledged that he would include me in meetings with big clients. 'Anything more than seven figures and I promise you, Junaid, you'll be in the meeting.' Some months later, he took me to lunch and, with a weary look, apologised for the fact that no such meetings had occurred. I assured him that he had not reneged on his promise – in that period, the company had failed to secure any such big-client meetings. His promise could be stated as follows:

Company has a potential deal worth seven figures
or more $\Rightarrow$ Junaid is in the client meeting

It was entirely fulfilled because, well, there was no way it couldn't be.

I came down hard on conspiracy theorists in the last chapter, so let me extend an olive branch here by suggesting there may occasionally be some logic behind their thinking. When social distancing measures were imposed during the Covid-19 pandemic, many people (and not just conspiracy theorists) flatly disregarded the rules. One may interpret their actions as reckless and selfish. Stated in terms of our $P \Rightarrow Q$ model:

Covid is a highly infectious and dangerous virus $\Rightarrow$ Mixing with others increases the likelihood of spread and is thus reckless and selfish

Nobody I know thinks they are reckless or selfish, yet most people are happy enough to accept this implication, as well as the conclusion. So we have a situation in which both Q and $P \Rightarrow Q$ are accepted as true. How does one then defend their rejection of social distancing without conceding they are reckless and selfish? Easy – by turning the above statement into a vacuous one. By downplaying the risks of Covid (for example, by comparing it to a common cold or dismissing it as a government hoax), the premise P – that Covid is infectious and dangerous – is deemed false. The implication is now valid by default; one may say, 'Sure, *if* Covid is as infectious and dangerous as you claim, my mixing would indeed render me reckless and selfish. But it isn't, so my actions are perfectly justified.'

This is an example of what psychologists term *motivated reasoning*, when we reverse-engineer beliefs to match desirable conclusions. In minimising Covid, conspiracy theorists can preserve their self-image of self-lessness and merrily continue with social interactions without the burden of guilt. Of course, one has to be prepared to dismiss the weight of scientific evidence that testifies to the threats posed by Covid – but the failure in thinking is not always strictly a failure of logic.

Proof by contradiction

We turn now to some of the techniques mathematicians employ to establish that something is true. One of the craftiest techniques is known as *proof by contradiction*, where one begins by assuming the literal opposite of a claim. It works as follows:

- Assume your claim (let's call it P) is false.
- Make a series of valid logical inferences until you reach a bare-faced falsehood (the contradiction).
- Assuming you were careful with each logical step, only one explanation remains; your initial assumption – that P is false – is mistaken.
- This means that P is true after all!

Proof by contradiction is an audacious line of argument – it forces you to indulge an opposing viewpoint and challenges you to examine its consequences and find a contradiction. To see the technique in action, let's prove that there is no largest number:

- First, assume the opposite – that there *is* a largest number. Call it L.
- Now add one to this number to get a new number, $L + 1$.
- $L + 1$ is larger than L, but L is supposed to be the largest of all numbers.
- This is a flat-out contradiction, so that initial assumption – that there is a largest number – is false.

Mathematicians have resorted to proof by contradiction to prove all manner of results, such as the fact that there are an infinite number of primes (though while it may logically demonstrate the existence of infinitely many primes, it does nothing to actually find them). And if you recall Cantor's argument (in Chapter 1) for why the real numbers cannot be enumerated in a list, that can be thought of as a proof by contradiction as well; if one starts by assuming the list can be created, then we can always find a number not contained in it.

You have almost certainly employed proof by contradiction in your everyday thinking. Whenever I misplace my keys, the first question I ask myself is whether they are still in the house. I can quickly convince myself that they are by arguing as follows:

- Assume the opposite – that I misplaced them while I was out for my daily walk.
- But since we always lock up when leaving the house, there's no way I could get back into the house without my own key.
- Yet here I am inside the house – a contradiction!
- Therefore, I must have had the keys with me, and they must be in the house (I could have left them in the keyhole, but I'm really not that thoughtless).

Arguing by contradiction can also be an effective rhetorical device. The next time someone asks you to 'hear them out' during a disagreement, it may be in your interest to do just that. If you adopt their view for a moment, you might then be able to demonstrate how it leads to an obvious falsehood, which would put them in a bind. Or, if you are in a romantic mood, you might use the same device to demonstrate your love for someone in the most robust way possible. The song 'If Ever I Would Leave You', from the musical *Camelot*, is a masterclass of proof by contradiction. Sir Lancelot grapples with his feelings for Guinevere and, as the title suggests, gives an account of what moving on from her would entail. The song proceeds season by season, with Lancelot concluding each time that he could not leave his beloved. Not in summer, with her 'hair streaked with sunlight', or in autumn ('I've seen how you sparkle when fall nips the air') or winter or spring. At the song's end, having exhausted all four seasons, Lancelot is left with the only logical conclusion: 'No, never could I leave you at all!' It goes further than the common sentiment of 'I can't imagine life without you'; Lancelot is imagining exactly that, only to demonstrate that it results in absurdity – his heart belongs to Guinevere after all.

Induction

The logic we've encountered so far in this chapter is of a *deductive* kind, where we start with a premise that we assume to be true and then, through valid inferences, arrive at other truths. It's a top-down approach to reasoning, whereby a general theory gives rise to specific consequences. However, logic works in the other direction, too, with specific observations leading us to general truths.

Here we will focus on one such approach: *inductive* reasoning. Mathematicians often call on it when they want to prove something about all positive whole numbers, which we'll refer to as n. Since there are an infinite number of cases, we cannot hope to check each of them in turn because we'd never reach the end – but the principle of mathematical induction saves us from having to do so. It works in two steps:

- First establish that the statement is true for the smallest value, $n = 1$ (the *base case*).
- Then show that whenever the statement holds for n, it also holds for the next value, $n + 1$ (the *inductive step*). In terms with which you will by now be familiar, this step demonstrates the implication: statement is true for $n \Rightarrow$ statement is true for $n + 1$.

Once both steps have been established, we can conclude that our statement is true for $n = 2$, $n = 3$, $n = 4$, and so on (each of these inferences, you might have noticed, is an instance of *modus ponens*). It's true for $n = 1{,}000$, a million, a billion, or any other whole number you care to choose. Dominoes provide the perfect analogy. To topple a row of dominoes, you need not resort to knocking down each individual domino. Instead, just give the first one a flick (the base case) and rely on the fact that each domino topples the next one along (the inductive step). The power of mathematical induction lies in its infinite scope; the inductive step can be applied indefinitely, akin to toppling an endless line of dominoes.

Let's use induction to prove one of my favourite mathematical results, which reveals an unexpected connection between odd numbers and squares:

> The sum of the first n positive
> odd numbers is equal to n^2.

Base case: the claim certainly holds for $n = 1$, because summing the first odd number gives 1, which is equal to 1^2.

Inductive step: the main idea behind this proof lies in how we can turn one square into a slightly larger one. Let's first consider a specific case, such as $n = 4$. The sum of the first four positive odd numbers is $1 + 3 + 5 + 7 = 16$, which is indeed 4^2 as the statement claims it would be. Here is the 4×4 square visualised as an array of dots:

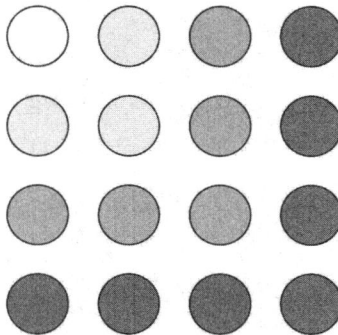

With the way the dots are shaded (in a reverse L-shape formation), you can see how each odd number corresponds to a 'layer' of the square. To get to the next square along – the 5 × 5 array with 25 dots – we just add the next layer, which corresponds to nine new dots:

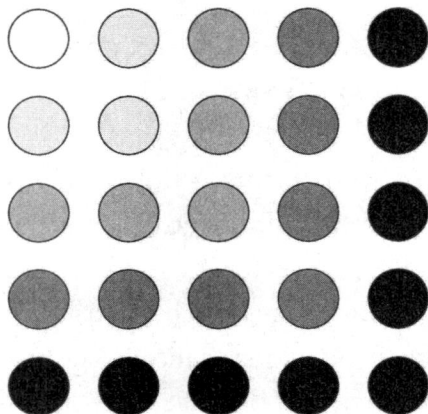

And this means that the sum of the first five integers is also a square number:

$$1 + 3 + 5 + 7 + 9 = 5^2$$

We now need to make essentially the same argument in a more general way, without assuming the actual value of n. If we assume our statement is true for *any value* of n, then we can say that summing the first n odd numbers gives n^2. Even without knowing the value of n, we can visualise an $n × n$ array of dots in much the same way as before (we use the ... notation to imagine this square stretching as far as needed, both horizontally and vertically):

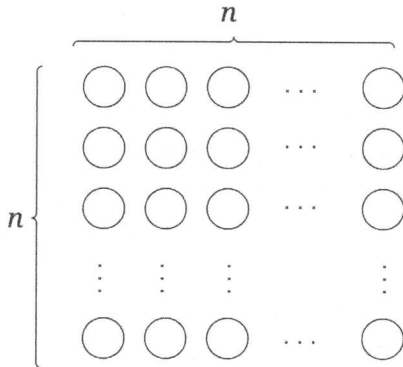

Just as we did previously, we now add the next layer to the square, giving a larger square with both dimensions measuring $n + 1$. As the following diagram shows, this amounts to adding $2n + 1$ new dots, which happens to be the $(n + 1)$th odd number.

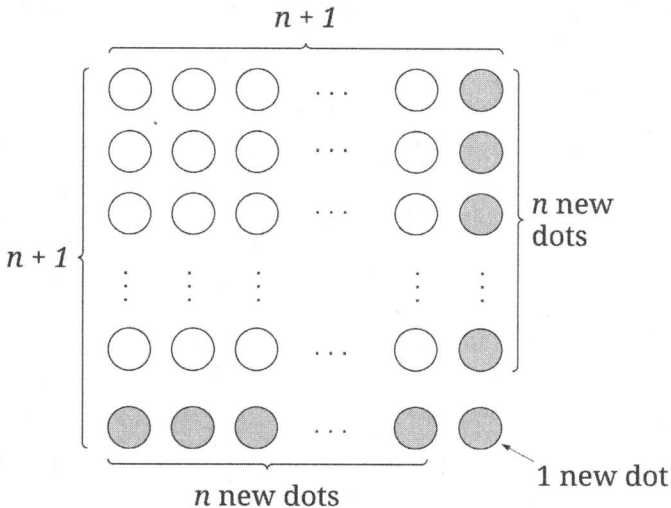

This is just what we wanted: summing the first $n + 1$ odd integers has resulted in the area of an $(n + 1) \times (n + 1)$ square, which is $(n + 1)^2$. So the statement is true for the value $n + 1$.

We have achieved the inductive step: whenever the statement holds for the first n odd numbers, it holds for the first $n + 1$ odd numbers. The principle of induction tells us that the statement is true for *all* values of n. Want to know the sum of the first thousand odd numbers? Easy: it's $1{,}000^2$ – a million.

Induction works spectacularly well for statements about whole numbers because once we have the inductive step in place, there is no scope for it to fail, even for extremely large numbers. There is a major distinction between this line of argument and one that simply presumes a pattern will hold up when we have observed it for a handful of cases. There are many plausible-looking mathematical patterns that hold up to very large values before failing (a trivial example: the statement 'n is less than a million' holds for values of n up to and including 999,999, before failing for the next number along).

In the real world, we rarely have the eternal certitude that numbers offer. I've included induction as a mental model precisely because of how often it is misapplied – the so-called *inductive fallacy* assumes that an observed pattern will continue indefinitely. This book's introduction began with the example of a love-addled mathematician who fell victim to the fallacy; in our new vernacular, he had the base case (loving his partner at the present moment) and, he claimed, the inductive step ('if I love her today, I'll love her tomorrow'). Chapter 1 gave one reason why this reasoning falls short: our emotions do not unfold in discrete steps. Induction does not apply to the continuum, so it gets us nowhere in situations that demand infinitesimally small units of analysis.

Induction may also break down simply because a pattern that has been consistent for a long time comes to an abrupt halt. The inductive step amounts to saying that incremental increases preserve behavioural patterns, yet on many occasions adding a new element disrupts the status quo. For instance, it may be true that adding a new member to your core friendship group increases your sense of collective belonging, but there are natural limits to social cohesion; there may come a tipping point where adding another member fundamentally alters its dynamic, to the detriment of your existing relationships. As an example, board games are greatly enriched by increasing the number of players, but things get awkward when the number exceeds the game's maximum player count.

When we attempt to topple an actual line of dominoes, there is always the possibility of failure due to lost energy in previous collisions, or because the next domino is angled incorrectly. The same can be true in our everyday lives; we may cling to an inductive logic that says we can maintain good habits (healthy eating, say), which proceeds like this: (a) I am eating well today (the base case); (b) if I'm eating well on any given day, then I'll eat well the following day too (the inductive step). The observation may be true for a while, until you reluctantly usher your three-year-old to yet another birthday party and find yourself confronted with the sight of rainbow sprinkle cake.

Sometimes it is circumstance that disrupts the rhythm of our day-to-day lives. For instance, our finances may abide by inductive logic for a period, as our savings accumulate interest with reassuring predictability. If I have a certain amount of funds in my account on a given day, I can be sure it will increase by a fixed percentage year on year. But if a medical emergency strikes and I have no choice but to withdraw funds from the account, my savings goals will suffer.

The complexity and unpredictability of the world makes induction a poisoned chalice. We may derive comfort in the certainty of everyday phenomena such as the setting and rising of the sun, but in the context of our lives the inductive fallacy has much in common with those productivity models that are predicated on perpetual growth (see Chapter 2). When we are on a hot streak, we can slip into the belief that the patterns of life will survive in perpetuity. Life hums to a more chaotic tune; love diminishes, habits break, income depletes.

Fuzzy logic

The logic we have covered so far restricts us to two options: a proposition may be true or it may be false, with nothing in between. This framing makes no room for *degrees* of truth.

Consider temperature. In our standard approach to logic, we can classify a given temperature as *hot* or *not hot* by specifying a threshold at which the classification switches from one to the other. Perhaps the cut-off is 20°C; anything at or above this temperature is *hot*, anything below is *not*. This is reminiscent of the step function from Chapter 1; a sudden jump at the 20°C mark means that 20.1°C is deemed hot whereas 19.9°C is not. One workaround, you will recall, is to 'smooth out' this function to remove the sudden jumps.

We might think of *hot* as having a 'truth value' of 1 and *not hot* having a truth value of 0. As shown in the following figure, every temperature can be assigned a 'truth value'. Really high temperatures have a value of 1 ('hot'), really low temperatures are given value of 0 ('not hot') and middling temperatures have a truth value somewhere in between. In the example shown, the truth value of 10°C is 0.33 and the truth value of 15°C is 0.65.

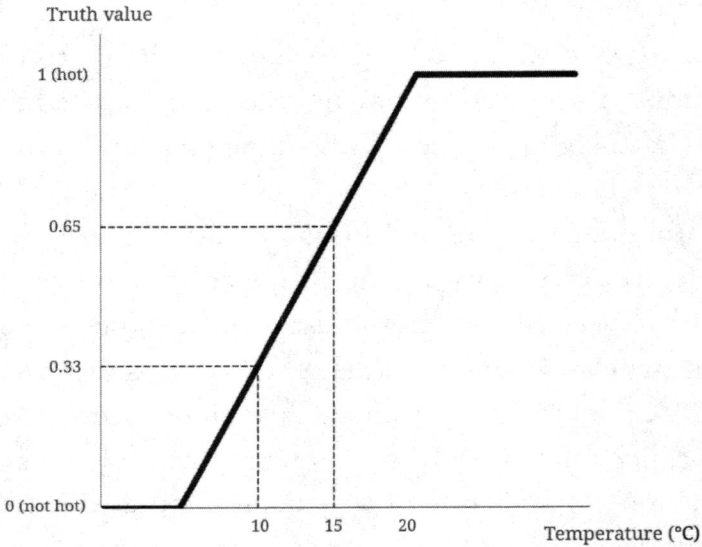

We might define a similar function for 'cold', 'cool', 'comfortable' or any number of other descriptors. So a temperature of 15°C may be 65 per cent hot (according to our 'hot' function), 15 per cent cold (according to our 'cold' function), 25 per cent cool and 60 per cent comfortable all at once. An air-conditioning system can then take appropriate action based on this collection of values. As more options are programmed, the system becomes more responsive to changing conditions and behaves more smoothly than a blunt system that simply turns on or off at a particular temperature.

This technique of assigning intermediate truth values, known as *fuzzy logic*, is a feature of many control systems (modern washing machines use it, for example, by analysing multiple factors such as soiling level and the presence of grease and soap concentration in order to balance a load more efficiently). More than that, it is another model through which we can shift away from binary thinking. Let's say you're recruiting for a role that requires a minimum of five years' experience and a master's degree. A traditional approach is to filter applicants based on *whether* they meet this criteria, but before placing a slightly less experienced candidate on the rejection pile, you might apply some fuzzy logic to evaluate the *extent* to which they meet each criteria. This offers a more textured picture of each candidate's credentials and highlights borderline cases; those who just fall short are still assigned a reasonably high truth value. If a candidate's profile is imbalanced – for instance, if they are very experienced but lack formal qualifications – then fuzzy logic provides a means of balancing both factors.

Where there are hard truths and clear binaries, the standard approach to logic serves us just fine. But when we feel constrained by black-and-white choices, fuzzy logic introduces some much-needed grey.

7

Combinatorics

Sizing up life's dizzying options

For a book showcasing the power of abstract maths, a chapter on counting may seem rather basic. Yet when it comes to everyday decision-making, the sheer volume and variety of our choices can be difficult to get a measure of. To be clear, this chapter does not deal with the infinite – we've covered quite enough of that in previous chapters. But even in the finite realm, there are many situations that require clever counting methods. It's not quite as easy as 1, 2, 3 but a trip to the ice-cream store will guide our thinking.

Permutations

Whenever I take my children out for ice cream, I'm reminded that one is never too young to experience decision paralysis. They will deliberate and argue as I look on, my thoughts inevitably turning to the mathematical question of how many options are on display.

Let's keep the example simple by supposing that the ice-cream vendor offers five flavours: chocolate (C), strawberry (S), vanilla (V), mango (M) and pistachio (P). We are allowed to select any three flavours (all different – no flavour can be chosen more than once). The ice cream is served in a cone, which means we must specify the order in which they are scooped; as my children will attest, the order is vital, depending on which flavour you want to savour right away and which you want mixed into the cone.

How many different triple-scoop cones are on offer? Assuming no flavour can be selected more than once (we'll deal with repetitions later), we have five choices for the bottom flavour. For each of those five choices, we have four remaining options for the middle, and then three choices for the top, for a total of $5 \times 4 \times 3 = 60$ variations.

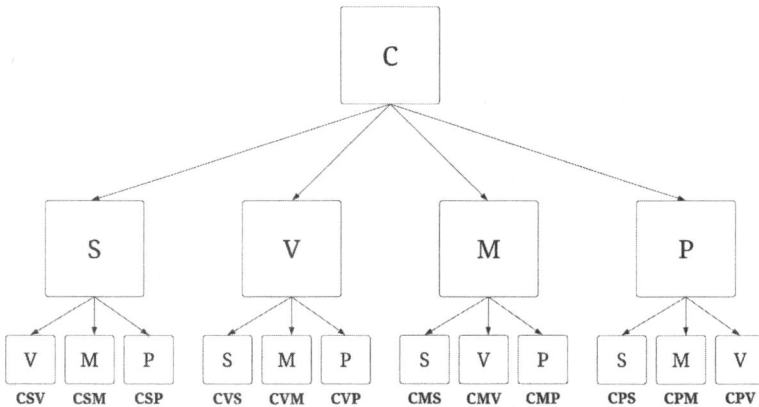

Here we see all three-flavour permutations where chocolate is selected first. For the next flavour there are four remaining choices, and for the third flavour there are three, for a total of $4 \times 3 = 12$ permutations. We can do the same for all five flavours, so the total number of permutations is $5 \times 12 = 60$.

A selection in which the order of the objects matters is known as a *permutation*. Counting the permutations can result in surprisingly large totals. Consider a simpler example where we have identified our three flavours of interest – chocolate, vanilla, strawberry – and want to try every possible arrangement. There are six in all: CSV, CVS, VCS, VSC, SVC, SCV. What about four flavours? It would be tedious to list them all, but following the same logic as the previous example, we know there are four options for the first flavour, three for the next and two for the third (with the fourth and final flavour then fully determined) – so a total of $4 \times 3 \times 2 \times 1 = 24$ permutations.

For five items, we have $5 \times 4 \times 3 \times 2 \times 1 = 120$ possibilities. In general, for N objects, we have $1 \times 2 \times 3 \times \ldots \times N$ arrangements. This is written as $N!$, where the exclamation, rather than an excited proclamation, denotes the *factorial* function and is shorthand for multiplying the integers between 1 and N.

The explosion of possibilities that arises with permutations has been seized upon by the choose-your-own-adventure genre in fiction. In 'Bandersnatch', a 2018 episode of the science fiction television series *Black Mirror*, viewers were able to make plot choices along the way, with each choice branching off towards a specific scene. This was before the age of generative AI, so the producers were not able to rely on an endless supply of content – in fact, every variation culminated in one of five story endings. Yet the show claimed to offer more than a trillion combinations, with a viewing time that ranged from forty minutes to two and a half hours depending on your choices. The eight-part Netflix heist drama *Kaleidoscope* opts for a similar approach with its episodes; viewers are invited to watch the first seven in any order of their choosing (only the finale is fixed), giving rise to more than 5,000 sequences. For a shuffling of just seven episodes, that is rich variety indeed.

When Apple's Steve Jobs launched the very first iPod and promised 'a thousand songs in your pocket', he may have missed a trick. Once you take into account all the permutations of a given selection of 1,000 songs, the possibilities amount to 1,000! – around $10^{2,567}$, comfortably more than the number of atoms in the universe. Even a modest playlist of ten songs offers over a million permutations; my running playlist is subject to experimentation, as I try to determine the optimal sequence of tracks (high-octane songs may feature at the start of a sprint session, or they might be deferred to the end of the playlist during longer runs).

Combinations

Back to the café, which has run out of cones and has resorted to serving ice cream in cups, with a material influence on the number of three-flavoured assortments. Intuitively, we might expect a smaller number of options, as each triple is now considered the same. When scooped alongside each other in a cup, the order of the flavours is of no relevance: chocolate–vanilla–strawberry is indistinguishable from vanilla–strawberry–chocolate and so on. However, since we already know there are six ways to arrange three flavours and, given that these will each be treated as the same, the total number of options is a sixth of what it was previously; hence, from a menu of five flavours, there are ten choices for the triple-flavoured cone.

This is a problem involving selection, where we simply pluck items off a list in any order, rather than a problem of arrangement. Each selection is referred to as a *combination*, very different from a permutation, where order is of the essence.

The agony of choice inflicted on my kids comes from the fact that there are many more than five flavours on the menu; at my last count, there were sixteen, which amounts to 560 possible three-flavour combinations. We can work out this total using formulas that generalise our earlier methods. A recurring theme in this chapter is that we can often count all the ways to make our choices without having to list them. Instead of the formulas, let me share one particularly elegant method that comes from an array of numbers known as Pascal's triangle, which has a 1 at its apex and abides by a simple rule: each row is bookended by 1s, and every number across the row is the sum of the two numbers above it. The second row thus comprises just the two 1s, but the third row has a middle entry of 2 (the sum of the two 1s in the row above), the fourth row contains a pair of 3s (the sum of 1 and 2), and so on. Here are the first eleven rows:

```
                        1
                     1     1
                  1     2     1
               1     3     3     1
            1     4     6     4     1
         1     5    10    10     5     1
      1     6    15    20    15     6     1
   1     7    21    35    35    21     7     1
1     8    28    56    70    56    28     8     1
1  9  36   84   126   126   84   36   9   1
1  10  45  120  210  252  210  120  45  10  1
```

Pascal's triangle is filled with patterns that could occupy us for hours on end. For now, we just need to draw on the convenient fact that to determine the number of ways to select three objects from a total of five, we simply find the value in the fifth row and third column (the only wrinkle is that we must adopt the convention popular among computer scientists that counting starts at 0 – the top row is the 0th row and the leftmost column is the 0th column). Trace your finger through the array and you should arrive at the number 10. The same thing will happen with any two numbers: if you want to know how many ways there are to select five objects from ten, simply locate the fifth column in the tenth row (not forgetting to count your rows and columns from 0); you'll arrive at 252 possible combinations.

The reason this method works is not immediately obvious, but I've spelled out some of the details in a footnote.* In practice, we'd just use the complicated formulas (which I've chosen to omit). Even so, I find comfort in the idea that the answer could be found simply by thumbing through Pascal's triangle. Want to know how many combinations of lottery numbers there are? Simple – in a format where six numbers are drawn from a ballot of fifty-nine, we just read off the sixth column of the fifty-ninth row of the triangle, to find a dispiriting figure of 45,057,474.

Combinations are one of my go-to models when I prepare my public talks on mathematics. As a speaker,

---

* The main idea is that, when choosing k objects from a total of N, we can consider a specific object among those N objects and see that our chosen collection of k objects either contains that specific object or it does not. In the first case, we are left to choose $k - 1$ objects from the remaining $N - 1$ objects (because one has already been chosen). In the second case, we have to choose k objects from the remaining $N - 1$ objects (because that object has been dismissed). We have therefore broken our problem into two smaller cases involving $N - 1$ objects, and get our total by summing the two values in those smaller cases. This is precisely the method of Pascal's triangle, in which we also generate new entries by summing two entries from a smaller case (namely, the row above).

I feel a responsibility to deliver a fresh presentation to each audience. At the same time, it seems unnecessary to build an entire talk from scratch each time. So how should I strike a balance between efficient preparation and novelty?

Suppose I have five talking points, each sufficiently distinct that any combination of three of them will comprise a legitimate presentation. By now we know that I have ten talk variations – one for each triplet. I sometimes pitch the five points to my audience and ask them to vote on their preferred topic, to democratically decide which three topics, and thus which of the ten talk variations, I will deliver. To the audience, I'm promising ten choices, yet I only need to prepare five sections and cobble together the selected three on demand.

The ratio of preparation effort to audience choice becomes more favourable as I insert more sections. Adding a sixth section requires marginally more effort (a 20 per cent increase in content), but I now have twenty talk combinations – a 100 per cent increase! Adding a seventh point increases the talk combinations to 35, an eighth makes 56, a ninth 84 and a tenth 120 – on Pascal's triangle, you'll find this sequence of numbers along a diagonal path. You can experiment with different variables, of course – for instance, you could increase or decrease the number of sections to be delivered. The point is that a simple application of combinatorial logic opens us up to more variety than we may otherwise have realised was possible.

The difference between permutations and combinations is easily muddled. The name of the 'combination locks' we rely on to safeguard our valued possessions is something of a misnomer because the order in which the numbers appear is fundamental – tapping the digits of your code in a jumbled arrangement won't get you far. Since the order matters, we might more accurately call it a 'permutation lock'. Just remember the following mantra: 'combinations for selection, permutations for arrangement'.

Many situations involve a mix of permutations and combinations, while some require deft manoeuvring to take advantage of both at once. Writing involves both selection and arrangement. Every sentence starts as a combinatorial problem: selecting the right words to convey an intended meaning. But a deliberate scrambling of these selected words can work to great effect, a rhetorical device is known as *hyperbaton*. 'The head that wears the crown lies uneasy' is a fine arrangement of words but switch a few around and you're left with the more memorable 'Uneasy lies the head that wears the crown'.

At the Tokyo Olympics in 2021, the failure of the US swimming team to win gold in the inaugural mixed-gender relay can be understood in terms of permutations. In this event, teams consist of two males and two females who, in some designated order, must swim the four standard relay strokes in a particular sequence: backstroke, breaststroke, butterfly and freestyle. For the team selectors, there are choices galore.

For a squad of twenty-eight swimmers, split evenly by gender, there are ninety-one ways to select two males (among the fourteen male swimmers) and ninety-one ways to select two females. Already we are at 91 × 91, or 8,281 possible teams. Next, for each possible team there are 4! (that is, 4 × 3 × 2 × 1 = 24) ways to line them up. In this very blunt analysis, a team can therefore line up in close to 200,000 ways. The reason the selectors do not suffer from decision paralysis is that the problem is laced with constraints; it is not the case that every swimmer can swim all four strokes competitively. For each of the four strokes, only a handful of swimmers are in contention.

As breaststroke is by some margin the slowest of the four strokes, conventional wisdom suggests assigning this leg to one of your two quicker male swimmers (unless you have supremely quick swimmers in the other strokes who can make up the deficit). For gold medallists Team GB, the order was female–male–male–female.[1] For silver medallists from China, male–male–female–female. In fact, seven of the eight teams that competed in the final opted for one of these two variants. The US opted for female–female–male–male, inexplicably pitting seventeen-year-old Lydia Jacobey against the formidable

Briton Adam Peaty in the breaststroke. In that leg alone, the US lost more than eight seconds to GB and, though they made up some ground in the final leg, they had to settle for fifth place.

The US learned the hard way that team selection is only one piece of the puzzle; team *arrangement* is just as significant.

Complexity

When my wife and I were expecting our second child, a friend gently warned me of the 'non-linear' nature of child-rearing (you can probably tell he was a mathematician). If parenting were a linear affair, each new child would induce a fixed new workload. It's natural to assume that the second child doubles the original parenting effort, a third child triples it and so on – already a daunting prospect, yet my friend's suggestion went further. He was a father of three and, as his offspring clung to various limbs, he explained how each child piles on the workload in unpredictable ways, meaning that having a second child means more than twice the workload. Can we put a number to his claim?

For the first five years of marriage, it was just the two of us. For all its ups, downs and in-betweens, there was a single relationship to manage within our household. Then along came our daughter Leena – a gift, a blessing and an incredible amount of work. As Leena gradually acquired sentience, new relational dynamics formed. Now that there were three of us, there were three relationships to manage: Leena and Mum, Leena and Dad, Mum and Dad.

Two years later we welcomed Elias, completing our family unit. Another gift, another blessing and, as my friend warned, a whole heap more work. It has been a thrill to witness the formation of new relationships, including a bond between siblings. Now there are four of us, meaning six relationships, and maintaining a happy, calm home environment is a daily exercise in diplomacy. The potential for discord – whether among adults, among the siblings, or between parent and child – is apparent as we strive to align preferences at mealtimes or discuss the best way to spend a Saturday.

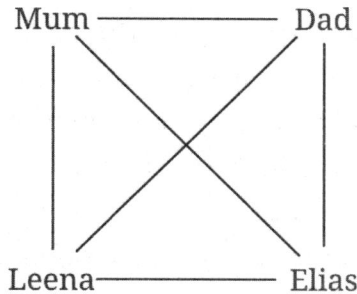

$$\text{Mum} \hspace{-0.5em}\begin{array}{c}\\\end{array}\hspace{-0.5em} \text{Dad}$$

Mum ————— Dad

Leena————— Elias

For my friend, the third child has extended the family network to ten relationships, a level of management I'm not sure my household could cope with. It may also explain the chaotic feeling of my own upbringing; as one of four siblings (the youngest, no less), I was part of a family dynamic comprising fifteen relationships, some of them more than a little sparky.

Let's examine these numbers a little more closely:

Number of people	2	3	4	5	6
Number of relationships	1	3	6	10	15

The non-linearity is apparent: the increase in the number of relationships itself increases by one more each time, because each new member forms a relationship with all the existing ones. Add a seventh member and they form six relationships, an eighth member forms seven new relationships and so on.

This number sequence might be familiar; it's one of the diagonals of Pascal's triangle (starting with the leftmost or rightmost entry of the third row from the top). The number of 'pairwise' connections is precisely the number of ways to select two objects from a larger collection. It is sometimes referred to as the sequence of *triangular numbers*, because each number can be visualised as a triangular array (much like *square numbers* can be for squares).

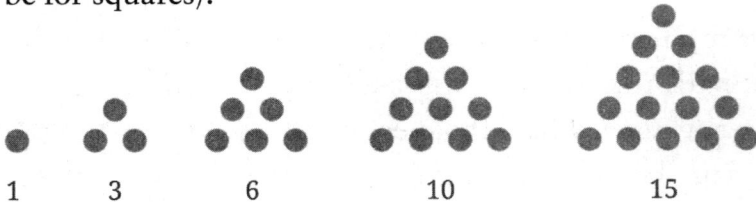

1   3   6   10   15

This pattern lends itself to a formula. For N people in a network, each individual forms a relationship with each of the remaining $N - 1$ people. Thus the total number of relationships in the network is $\frac{N \times (N-1)}{2}$ (we have to divide by two to avoid counting each relationship twice; once we've counted a relationship between Alice and Bob, we can ignore the relationship between Bob and Alice).

The expression $\frac{N \times (N-1)}{2}$ can be written as $\frac{1}{2}(N^2 - N)$. The appearance of N^2 confirms that the growth is quadratic, rather than linear. Roughly speaking, this means that if the number of people in a network doubles, the number of relationships will increase by a factor of around four; my friend was onto something when he advised caution around increasing the size of our family.[2]

In a parenting context, I hope it's clear that I'm being slightly tongue in cheek. The workplace is a different matter. In fact, it may explain the tendency of many organisations to structure themselves around small teams. Jeff Bezos famously advised companies to avoid having teams that could not comfortably share two pizzas – I wonder if he knew how quickly the number of pairwise connections between team members grows. Teams that exceed the 'two-pizza limit', according to this theory, fall into trappings such as 'social loafing', where members put in less effort as part of a group than when they're working alone; for managers, it becomes nigh on impossible to ensure that team members are contributing equally and to maintain a consistent work culture.

The complexity of a group extends beyond pairwise relationships. Schoolchildren, with their various friendship groups and playground politics, demonstrate this better than anyone. Teachers typically must contend with class sizes of thirty or more (in many classrooms I have visited around the world, groups can number more than a hundred), which results in some 435 relationship pairs among the students ($\frac{30 \times 29}{2}$). But if we consider groups of *any* size – three classmates, four classmates and so on – the number of possibilities quite literally compounds. As we saw in Chapter 4, when forming a group (or a *subset*, to use the proper term), for every one of the thirty students there are two choices: they are either in the subset or not. With thirty binary choices to make, we can form 2^{30} different subsets. There is a subset containing every student in the class and another containing no students, with every other subset (more than a billion of them) containing at least one student. And if a new student joins the class, the number of subsets will automatically double (to 2^{31}).

The next time you find yourself nodding along to critiques of our education system, pause for a moment to appreciate the sheer complexity teachers are faced with. The possibilities don't even end here: as well as managing an unfathomable number of relationships in any classroom, for any given assignment teachers must determine how best to split up their students. Every student must be allocated a group and no student can be allocated more than one. This, you will recall from previous chapters, is what mathematicians call a *partition*. But how many possible partitions exist for any given group size? For a group of three students (Alice, Bob, Charlie) there are five possibilities:

- Alice–Bob–Charlie
- Alice–Bob, Charlie
- Alice–Charlie, Bob
- Bob–Charlie, Alice
- Alice, Bob, Charlie

With a group of four, there are fifteen possible partitions, while a fifth group member would take us up to fifty-two partitions and a sixth to 203. This sequence grows fast – again on the order of exponential growth. A mere twelve students give rise to more than a million possible partitions.

I suspect the maths of partitions does not feature much in teacher training courses, which is probably for the best. But perhaps a memo should be sent to those policymakers and critics who choose to make simplifying assumptions around classroom management. As a parent, I want my children's individual needs to be attended to, but I have sympathy for teachers who are confronted with a combinatorial explosion of possibilities and might justifiably resort to more standardised practices such as seating arrangements. Two pizzas will never be enough to feed a class of thirty, after all.

Entropy

During my successful run of appearances on the television game show *Countdown* in 2008, one sign of my misspent youth became very apparent. For the uninitiated, *Countdown* consists of 30-second number and letter rounds that culminate in the 'Conundrum', in which contestants must unscramble a nine-letter anagram. If the letters are all distinct, there are 9! possible arrangements – which, you'll remember, means $1 \times 2 \times 3 \times 4 \times 5 \times 6 \times 7 \times 8 \times 9$, or around 360,000, only one of which corresponds to an actual word.

One case we haven't yet covered is repetitions. Suppose that the string of nine letters contains some duplicates – say there are three occurrences of the letter E. Intuitively, we'd expect the number of possible arrangements to fall. Let's make this more precise: for each arrangement of nine letters, of which three are E, there would be six times as many arrangements if those three letters were distinct (this is because there are six ways of arranging those three letters: $3 \times 2 \times 1$). So the total number of nine-letter arrangements, with three Es and six other distinct letters, is reduced by a factor of six, but still presents a daunting 60,000 possibilities.

At first thought, 30 seconds seems a pitifully short amount of time to extract one word from a crop of 60,000, let alone 360,000, yet seasoned *Countdown* players regularly solve the Conundrum in the allotted time, and the best usually require no more than a couple of seconds. Their method is not to trawl through the thousands of options. An anagrammer applies a range of shortcuts: they may focus their search on words with common prefixes (e.g. MIS-, PRE-, DIS-), or suffixes (e.g. -ING, -IOUS, -IEST), or visualise the word's structure by separating out the vowels (four successive consonants are rare). So, if I give you the scramble DENTEPERV (which appeared during my run on the show), there is nothing stopping you from unravelling the solution before too long.*

The effort to reduce the space of plausible solutions is based on the idea of *entropy*. This is one of those terms that recurs across science and relates to the degree of disorder or uncertainty in a system. The second law of thermodynamics says that when a system is left alone, its entropy increases. It's why a tidy office is bound to regress to a messy state, why ice cubes melt at room

---

* The answer is ... PREVENTED.

temperature and why our body's cells decay as we age. This concept postulates that, over time, entropy will eventually become so high that no energy is available. It explains why most businesses die with a whimper: unchecked disorder leads to inertia, and an inability to get anything done.[3]

In maths, entropy is a property of information. Its mathematical usage was coined in 1948 by Claude Shannon, who sought a term to capture the idea that the more surprising a message is, the higher value that piece of information has. On the advice of the polymath John von Neumann, Shannon opted for entropy, partly because of its related use in thermodynamics, and partly because, as von Neumann said, 'No one really knows what entropy really is, so in a debate you'll always have the advantage.'[4]

Entropy is best understood through examples. The outcome of a die roll, from the perspective of someone who can't see it, has a higher entropy than a coin flip because the former has six equally probable outcomes, while the latter has just two. A coin landing on heads (or tails) is objectively less surprising than a die landing on any particular number – there was a one in two chance of the first event happening, compared to a one in six chance of the latter.

For a more extreme example, the standard 3 × 3 × 3 Rubik's Cube has more than 43 quintillion (that's 43 billion billion) possible configurations. If you had to guess the random configuration of the cube that's sitting on my desk, your degree of uncertainty would be unfathomably high – the situation is highly entropic. But if I told you I was a single twist away from solving my cube, you would be down to twelve possibilities. As Rubik's Cube solvers home in on the correct solution, they are drastically reducing the number of possible configurations that their cube can take on. For an observer who cannot see the cube, the more solved it is, the less uncertainty surrounds the configuration – and the lower the entropy.

Playing board games is often a prolonged act of entropy reduction, in which maximising your information gains is key. In *Guess Who?*, a competitor is faced with pictures of twenty-four characters and has to determine their opponent's choice by asking yes/no questions. The game is, in effect, a race to zero entropy: by the time you've pinned down the candidate, you've removed all the uncertainty. The highest-risk, highest-reward strategy is to take a punt on a specific character with your first question ('Is it Bernard?'). If your opponent does indeed have Bernard in their possession, you win right away. But that is a mighty *if*: this strategy will only work one out of every twenty-four times.* In all the remaining cases, you'll be lumped with twenty-three faces to choose from. A much

---

* Strictly speaking, there are twenty-three candidates because you can rule out your own character.

better strategy – in fact, the optimal one – is to ask a question whose answer divides the candidates into two equal groups. If twelve wear a hat and twelve do not, asking: 'Does your person wear a hat?' is guaranteed to leave you with half the number of candidates. It will take you no more than five questions to identify your opponent's character from the original batch (after three halvings you're down to three candidates, from which point you can do no better than taking a random guess each time). Each question rewards you with some gain in information; the best questions are the ones that afford you the largest gains.

The game that Claude Shannon actually had his sights on when describing information entropy was chess,[5] whose possibilities once more take us beyond atoms-in-a-universe comparisons. Consider just the first two moves in a game: each player can make twenty moves, giving rise to $20 \times 20 = 400$ possibilities. Assuming each subsequent move carries forty new options, this leads to Shannon's estimate of 10^{120} possible games. Each board position carries a degree of uncertainty in knowing which move comes next. In a position of high entropy, there are many credible moves, making it harder to predict how the game will unfold. But if a player can establish positions they're familiar with and restrict their opponent's moves, they can more reliably anticipate upcoming exchanges. In a game with enormous complexity and uncertainty, players are striving to manage the entropy of each position, while forcing their opponent into a position of zero entropy. It's at this point that they can be certain of their opponent's next move; a checkmate will often follow.

There are two factors in play when we're solving games or puzzles: the space of all permutations, and the methods we have available to search through them. Despite its intimidating number of configurations, the Rubik's Cube is readily tamed by the many solution algorithms now known to us.* A complete beginner can learn how to solve the cube from any starting position over the course of an afternoon. However, for 'speedcubers', the thrall of a Rubik's Cube does not come from the mere act of solving it, but from finding more efficient algorithms that shave seconds off their personal bests (or, at the level of world-class competitors who can solve the cube in under five seconds, *milliseconds*). For a mathematician, there's satisfaction in figuring out the smallest number of moves needed to solve the cube from any given starting configuration; it turns out that a solution is always available within just twenty moves.

With its incredible complexity, chess cannot be *fully* solved in the same manner – but it can be mastered. Part of Shannon's motivation was to understand how the

---

* Thanks largely to the algebraic structures inherent in the cube, which allow us to bring to bear ideas from a field of maths called group theory to devise such routines.

ideas of entropy might inform an approach to designing chess-playing machines. Decades later, advances in AI have reached a point where even relatively modest computer programs can now beat the world's most skilled grandmasters. Yet chess remains a creative endeavour for humans. Limited as we are in our processing capabilities relative to a computer, we are constantly having to seek out strategies for entropy reduction. The appeal of anagrams, similarly, is in no way diminished by computers' ability to exhaustively search through all possible options in an instant. In the absence of an all-encompassing algorithm fit for the human mind, we are left to devise our own shortcuts and appeal to our intuitions.

During my marathon board-gaming sessions, which take place in my highly disordered home office, I occasionally reflect on how problem-solving inverts the natural order of things. Games and puzzles give us practice in bringing order and predictability to complex situations where uncertainty would otherwise reign.

Entropy reduction certainly pays dividends in the real world. A business that manages to reduce uncertainty around its customers' behaviour is better placed to reach them with targeted advertising. When a company launches a new product, it faces a lot of uncertainty. Within its customer base, there are those who will buy the product and those who won't. What the business wants is to identify an attribute that accurately sorts its customers into likely buyers and non-buyers. If the company is selling pensions, for instance, it may be guided by age, with customers above a certain threshold marked as likely buyers and younger customers marked as likely non-buyers. Now that the customers have been split into two groups – older and younger – the business can offer the pension to the first group and ignore the second, with greater confidence in how each customer segment will respond compared to a campaign that targets all customers at the same time. This case is straightforward because age fits so neatly with pensions. In general, companies need to uncover the attributes that divide their customers so each resulting smaller group is more predictable than the customer base as a whole.

Efforts to reduce entropy can be observed everywhere. To an investor, uncertainty represents risk; entropy reduction merely amounts to the sensible practice of balancing risk and return. In an educational context, entropy may correspond to our degree of ignorance in a subject, and we may direct our learning to topics that offer maximum information gains. To an author, entropy may relate to a plot; a well-crafted narrative will seamlessly transition from an initial state of high entropy, when we are introduced to the fictional world, its characters and their motivations, to one of low entropy, when all the subplots come together in a satisfying, coherent conclusion.

But there are also situations in which we might welcome uncertainty. Doesn't spontaneity make life interesting? Innovation depends on allowing ourselves to be led down uncharted paths. Not every career path needs to be mapped out fully. Love, meanwhile, is hardly reducible to entropy-reducing algorithms – after all, surprise and novelty are the drumbeat of a budding romance. Every meaningful relationship requires some yielding to emotions and intuitions that are less easily codified by the tools of information theory.

The story of our lives is marked with unknowns. It can be reassuring to manage the unknowns, but it can also be liberating to let the story play out and embrace the surprises.

Game trees

One model for understanding how we make our life choices is the game tree. Let's start with the game of noughts and crosses. At any moment, one of the two players is faced with a finite collection of possible moves. Each of those moves branches into another collection of moves, giving rise to a tree-like structure (or, to be more accurate, an upside-down tree). The points on the tree (the board positions) are called nodes. Here is a snapshot of a game tree that shows the two moves available to O after X has placed their first move in the middle (there are, of course, eight possible moves, but thanks to symmetry we can treat the four corners as the same, and likewise the side squares – so the only meaningful choice is corner versus square):

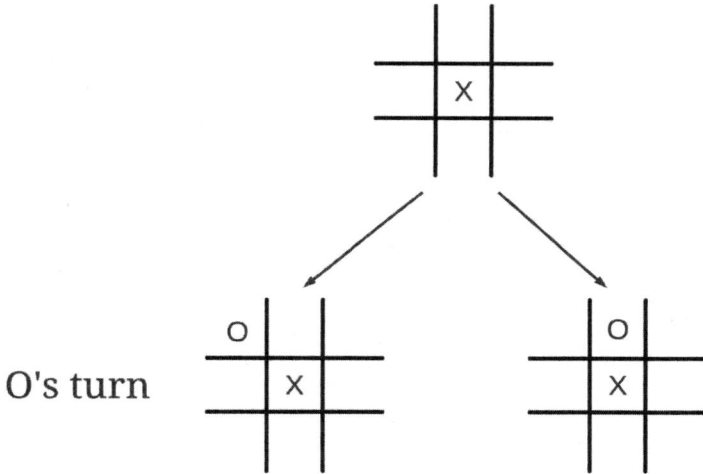

O's turn

To determine which of these moves is more likely to result in a win for O (or less likely to result in a loss), we need to decide on a search path. We could head down a single branch, working through several layers of the tree, node after node. In other words, we would select one of the available two moves and imagine a specific scenario in which X responds to that move. If this sequence of imagined moves results in a board position that favours O, the original move is a good bet. If it doesn't, we can try the same thing along a different branch. Here is an expanded game tree, where O has opted for the side square:

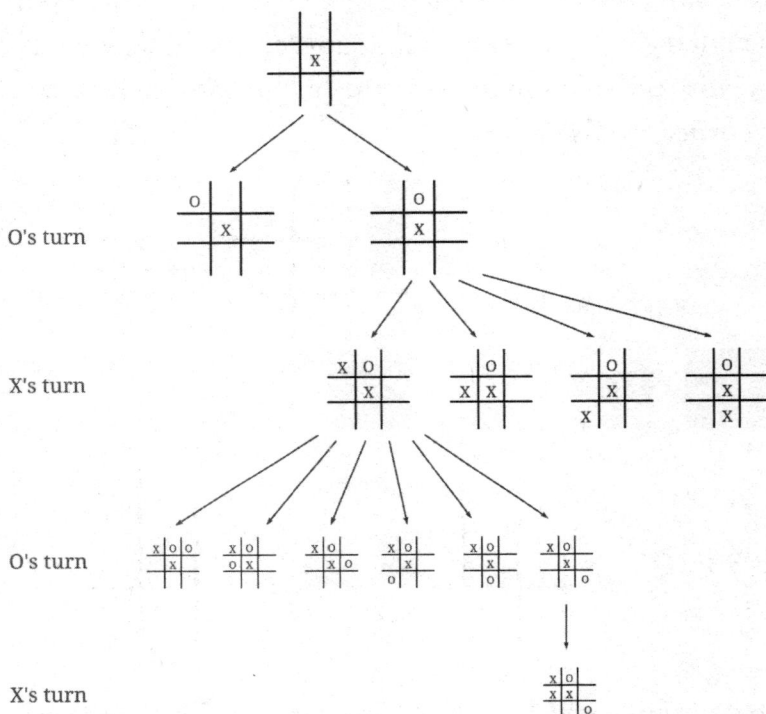

By studying this branch, you should be convinced that the side square is a terrible choice for O if X places in the middle. We could continue this way down other branches – noughts and crosses is a sufficiently simple game that we could plausibly map out the entire tree. That also renders it a fairly dull game (in which, it turns out, both players can easily avoid defeat every time with optimal play).

Now imagine the game is chess, where we have a staggering array of nodes, each representing a possible future for that game, that no player (whether human or computer) can be expected to work through in full. What's a chess player to do?

The first option is to venture down a single branch of a game tree to see if a particular move plays to your advantage. Due to limits on your memory (not to mention your time), it is not possible to venture down every path with chess – there are too many and, within each branch, the possibilities compound. You might instead sample as many branches as you can, compromising on how far you traverse each path – looking ahead, say, a couple of moves at a time.

These two approaches are called depth-first search and breadth-first search, respectively, and they represent a trade-off in situations in which exploring every possibility is impractical. As you've surely anticipated, they don't just apply to chess or games. In life, we are all constantly fighting this tension between depth and breadth. When pursuing the perfect career, for instance, we often have to decide between a prolonged tenure at a single organisation – where we might have opportunities for promotion into more senior roles – and multiple shorter stints across different organisations, which might expose us to a richer variety of people and skills.

There is no objectively superior approach here, no absolute answer on whether you should focus on existing opportunities or direct your efforts towards new ones. A depth-first search pays the ultimate dividend if the jackpot is lurking down that path, but there's always the possibility that true fulfilment lies down a path you have yet to pursue. We simply do not have time to realise all our future hopes and dreams, but being aware of the trade-off between depth and breadth can help set the direction of our future travel.

Another key element of game trees is how we should evaluate the strength of each node. With chess, we glossed over the question of what is meant by a 'good board position'. There are many ways to define this, from crudely counting how many pieces you have on the board relative to your opponent to more sophisticated measures that take into account your pieces' positions. Each board position can then be assigned a number using an 'evaluation function', with a higher value corresponding to a stronger position.

We still need a method of deciding which move to opt for among the choices we're faced with – a decision that should anticipate what our opponent is likely to do. A popular option is the *minimax* algorithm. Suppose you are in the midst of a two-player game (it doesn't have to be chess) and it's your turn. To keep the example simple, let's further suppose you are considering two options for your next move, which we'll refer to as A and B. For each of those, your opponent can respond in two different ways, which amounts to four different board positions after *their* next move. Using your evaluation function, you can assign a value to every possible move; higher values are favourable to you. Perhaps our simplified game tree looks like this:

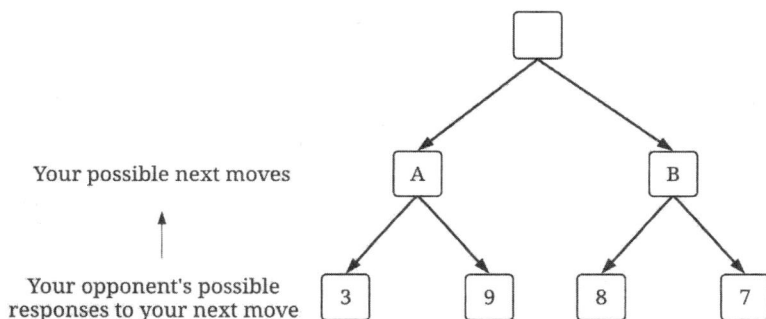

Your possible next moves

Your opponent's possible responses to your next move

You can now work backwards from your opponent's next move, examining each of their four possible responses in turn. The minimax algorithm presumes that your opponent will make the best move available to them. In other words, they'll choose the option where the board position has the *smallest* value. In this example, if you were to execute option A, your opponent can select between two moves, the first of which results in a value of 3 and the other a value of 9. Since higher values are better for you, they will choose the first of these moves. Similarly, if you opt for B, your opponent will choose the move that results in a value of 7 over the one that has a value of 8. So option A leaves you with a value of 3 by the time your opponent has responded, while option B leaves you with a value of 7. You should therefore opt for B over A.

Minimax is so named because it minimises the *maximum damage* that your opponent can do to you. It is the mathematised 'lesser of two evils' option that we often find ourselves deploying in real life. It's the political candidate who, for all their flaws, is judged to inflict the least harm. It's the traffic route that, however undesirable, results in the shortest delay among all available routes.

Minimax confronts head on the tough choices life presents to us. But if, at any stage, we feel discontent with a handful of underwhelming options, we may just need to reconfigure our search and take a chance on unexplored branches. You can always spoil your ballot paper, or avoid traffic altogether by changing your destination.

8

Dimensionality

Escaping our perceptual limits

My political identity is hard to pin down. I've voted for three of the major UK political parties at general elections and have given the nod to candidates from other parties in local contests. Politics did not feature very much in my upbringing – it was only at university that I realised being a 'leftie' can refer to something other than the hand one writes with.

The left–right terminology has its origins in the seating arrangements of the post-revolutionary French parliament. Radicals wishing to topple the king occupied seats on the left of the assembly, with their aristocratic peers on the right. The notion that our collection of political beliefs and dispositions can be reduced to a single point on a continuum is now deeply entrenched. The spectrum of political beliefs is often presented as precisely that – a continuum in which socialism and communist ideals are situated on the left, with reactionary ones on the right. Liberals are slightly to the left of centre, mirrored by conservatives on the opposing side. The 'centrists' that occupy the middle of this line apparently represent some kind of compromise between the two.

This renders me a floating voter, moving from one part of the left–right continuum to another as my party of preference changes. Yet my overarching political beliefs are stable, so any framework that purports to characterise them should keep them fixed in place. The problem with the left–right continuum is that wherever I land, some aspect of my political outlook will be ignored. Some elements of both left-leaning and right-leaning politics appeal to me, while there are others that I bristle at. As a recipient of free life-saving healthcare and life-enriching education, I am a passionate supporter of the welfare state. But before calling me a socialist, you should also know that some conservative values, such as the importance of traditional family structures and religious traditions, align with my own. The 'left' and 'right' labels are not as diametrically opposed as the continuum model suggests, nor can they be neatly 'averaged' into a centrist position. And it's not just the labels that are at fault, but the presumption that the intricacies of our underlying beliefs can be mapped onto a one-dimensional line. A continuum allows for an infinite number of possibilities (uncountably many at that), though that's of little comfort to a voter unable to place their own identity among all those options.

A common workaround is to add another line, or axis. Political psychologists have suggested, for instance, that our beliefs can be divided into two distinct spectra: the economic (covering views on issues such as taxation and wealth redistribution) and the social (covering policy areas such as abortion and religious teaching in schools).[1] This construct allows for a broader range of opinions. My first boss was a paid-up member of the US Republican Party, identifying himself as 'socially liberal and fiscally conservative'. There is space for him on this biaxial plane, towards the bottom right. There is room, too, for someone like me, whose leanings are somewhat the reverse of this; my views place me somewhere in the top left quadrant.

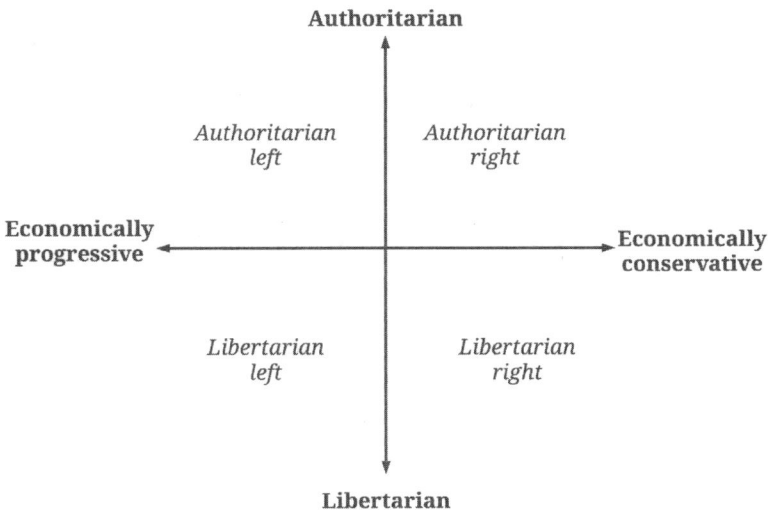

Yet the model leaves me feeling queasy at the suggestion that my views tend towards the authoritarian end of that vertical spectrum. It also feels simplistic in its separation of economic and social policies; in reality, the two are entwined. A champion of social equality is likely to feel solidarity with labour rights advocates; debates regarding access to education and healthcare, similarly, have an economic element, as policymakers reckon with the trade-offs of resource allocation. And libertarian policies that seek to diminish the role of government are deeply implicated in wealth disparities between rich and poor.

Because two axes still seem inadequate, adding a third is a natural next step. You may have a vague sense of where your views lie on this three-dimensional chart, but you might once again feel that it neglects an essential aspect of your worldview by restricting your choices within such tightly defined parameters.

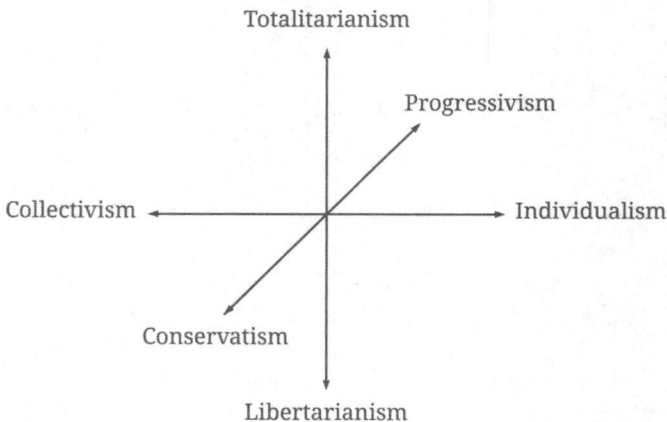

These models have the attractive quality that they are easy to visualise and digest (although many people, myself included, have trouble visualising three-dimensional space in their minds). They inflict a cost, however, of limiting our understanding of a situation to a very small number of factors – what I'll call *low-dimensional thinking*. Furthermore, it is not always clear if the specified dimensions are as distinct as we are led to believe.

Mathematics is unlikely to resolve the complexities contained in our sociopolitical views. What it can do is bring clarity to the concept of *dimension*, so that when we attempt to classify our beliefs – along a continuum, along two axes, or however else – we are more aware of our simplifying assumptions.

A major advantage of the mathematical conception of dimensionality is that it is not confined to the one, two or even three dimensions that we can visualise. Dimensionality is not constrained to our sensory perceptions – mathematics gives us a way to think in as many directions as we need to.

Flatland

It takes a leap of imagination to contemplate life outside of three dimensions. Before we step up to the fourth dimension and beyond, let's ask what life would look like in a mere two dimensions. This is the world depicted in *Flatland*, a nineteenth-century novella by Edwin Abbott Abbott.

Flatland is told from the perspective of *A. Square*, a respected four-sided denizen of a two-dimensional world. Bound within a flat plane, the occupants of Flatland perceive one another as nothing more than lines. Due to the foggy atmosphere, objects become brighter as they approach, giving some perception of depth and allowing a triangle, say, to be distinguished from a pentagon. The shapes possess no notion of life beyond their two-dimensional reality.

Flatland served as a satirical commentary on the self-imposed limits of social perspective in Victorian England. Abbott was a headmaster and clergyman, and through his penchant for geometry he called foul on rigid class structures, the marginalisation of women and religious dogma. In Abbott's imagined world, the more sides a shape has, the higher its social standing. The circle (an 'infinitely-sided' polygon) represents the priestly ruling class. Women, on the other hand, are lines, the lowest of all casts. Viewed end-on, they appear as points and risk impaling their male counterparts. For this reason, they are required to maintain a 'peace cry' to notify others of their presence.

One day, the square is visited by a sphere from Spaceland, a world that we would recognise as existing in three dimensions. But the square is not capable of seeing the sphere as three-dimensional. Whereas you and I can imagine the sphere hovering over Flatland and sinking through it, the square only experiences the sphere as a series of circular cross sections.

Understandably perplexed, the square is lifted from its environment and given a tour of the sphere's three-dimensional world. It hovers above its previous home, seeing everything from atop for the first time (including the insides of its cohabitants). Enlightened by his three-dimensional worldview, the square now wonders what life might look like in four dimensions; at this, the sphere, affronted by such a notion, banishes the square back to Flatland.

The major lesson of *Flatland* is that we struggle to perceive realities beyond our immediate sensory experience. As much as we may pass judgement on those who exhibit the limitations of a two-dimensional worldview, we should spare some pity for those who are unable to conceive of a world beyond three dimensions – and we may extend that pity to ourselves.

Is there a hatch through which we can escape our sensory confines? Plato certainly thought there was. Flatland resembles the shadowed walls of the Greek philosopher's allegorical caves; Plato likens 'untutored' people to prisoners chained inside a cave, unable to move their heads. The prisoners' eyes are fixed on the cave wall. Behind them, a fire burns and puppeteers hold up figures that cast shadows on the walls. The prisoners can only see the shadows – the restrictions imposed on them have deprived their senses of an entire dimension.

Plato suggests that we all experience some degree of blindness; our words, he says, refer not to the physical objects we see but to hidden 'forms' that are access-ible only to our minds. The item that you hold in your hand, for instance, is a physical manifestation of the idea of a *book*; a shadowy projection of a true form that completely represents all that a book can be.

Mathematics itself is also viewed by Plato in terms of forms. Take the circle from *Flatland*: a straightforward object that we can visualise and draw with little effort. Yet nowhere in the physical universe will you find a perfect example. The World Freehand Circle Champion (yes, there is such a thing) may have a strong claim to such an object, but even theirs is an inexact rendering. Despite this, mathematicians have, over the course of millennia, established a deep understanding of the idealised circle, from formulas for calculating its area and circumference to the starring role played by the mathematical constant π and much else. There is such a thing as a circle, but the only means of grasping its true form is through abstract, intellectual endeavour.

Plato is an optimist; he notes that upon their release, the prisoners – much like Abbot's square roaming around three-dimensional space in a state of enlightenment

– realise the error of their ways. Plato asks how the rest of us may similarly achieve this enlightenment and acquire a firmer grasp of nature's true forms. For Plato, this is the foremost challenge of education. To become *tutored* is to gain a richer understanding of the world we inhabit by defying the limits imposed on us by our perceptions.

Flatland was published during a period of rampant scientific innovation. As physicists experimented with new discoveries such as electricity, they realised that points in space could be described not just in terms of their three positional values, but also in additional variables such as the magnitude and direction of electromagnetic fields. The standard three dimensions of space would no longer suffice.

Soon after *Flatland*, Einstein's notion of *spacetime* was predicated on a four-dimensional view of the world – he took the three familiar dimensions of space and wove time into them. Spacetime tells us that our perception of time depends on our speed and the gravitational field we are in – our three-dimensional experiences are, in a sense, 'shadows' of four-dimensional spacetime. Einstein saw further than most, but reality may not end at the four dimensions he conceived. Current versions of string theory invoke up to eleven dimensions.

Grasping our universe means going beyond even the perceptual limits of Einstein's theories, just as grasping political beliefs requires more than a handful of characteristics. It would be useful to have a framework for dimensionality that allows us to keep on adding new dimensions – and the mathematical concept offers precisely that.

What is a dimension?

Our familiar notions of space and direction get us to three dimensions: a point has zero dimensions, a line has one, a square has two (length and width) and a cube has three (length, width and height). We can think of each dimension as building on the previous one: a line is the path traced by a point as it moves along in a single direction. If the line then moves along in a new direction, it traces a square. When the square travels in yet another new direction, we end up with a cube. At this point, things get tricky because there is no obvious 'new direction' for the cube to travel in.

Recall how the square in *Flatland* perceives the three-dimensional sphere as a series of two-dimensional circular cross sections. If we raise this example by a dimension, we would experience a four-dimensional object as a series of three-dimensional cross sections, which we can just about visualise.

Think about how we draw a cube on a piece of paper. The paper is flat and cannot possibly capture the entire cube – it does not show all six faces at once or show any

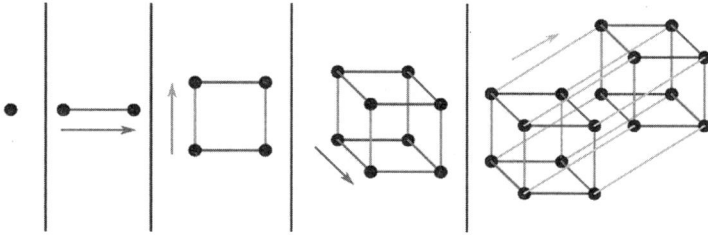

Visualising up to four dimensions. Each new dimension
arises from tracing the previous object along a new direction.
The point traces a line, the line traces a cube and the
cube traces a hypercube.

actual depth – but it serves as a two-dimensional projection that gives us enough information to infer the cube's structure. So just as an ordinary cube consists of two-dimensional square faces, a 4D hypercube will consist of faces that are each three-dimensional cubes.

This kind of visualisation is tough going, and that's before we attempt a leap into the fifth dimension and beyond. But now that we have some tangible mechanism for how we can transition from one to two to three and even four dimensions, we can keep going without the need to picture each subsequent iteration. If we can find a way to properly define what we mean by a 'new direction', we can apply our process over and over to generate however many dimensions we desire. This is the essence of abstraction, where the objects we visualise are just specific, low-dimensional instances of a general concept.

The first thing we need is a notion of zero; an origin from which we can get our bearings. We also need vectors – lines that connect our origin to other points in the space, each one corresponding to a specific direction. We now ask how many vectors we need to describe every point in that space.

In the case of a line, every point can be described with respect to a single vector. Think of a number line and take any non-zero value away from zero, such as 1; our chosen vector is thus the segment of line that connects 0 and 1. To reach any given point on the number line, we travel some amount along the direction of our vector. To reach the number 2 we travel twice the length of our chosen vector; to reach 17, we travel seventeen times its length. We can use fractional amounts to get to points that are closer to 0 (for instance, we travel half the amount to reach 0.5). We can even describe negative numbers by moving in the opposite direction – we end up at the number −5, for instance, if we move five times along our chosen line segment, *left* rather than right.

If I consider my house to be 'point zero' on my street and choose my neighbour Josh's house as a reference, I can describe the position of every other house on the street as a multiple of the vector connecting my house to Josh's. It may feel convoluted, but it can be done in principle (house numbers are essentially a variation on this idea, with the start of the street taken to be the origin).

In a flat plane, reaching every point requires a second reference vector. To direct a tourist in my neighbourhood to the local coffee shop, I cannot simply point them along the vector that passes through Josh's house. Our street, even if extended in both directions, does not house any such café (alas); a second vector is required.

In navigational terms, the two vectors of a flat plane are usually described in terms of a line pointing north and another pointing east. Any point in this space can be reached as some combination of the two (with negative amounts used to reach points to the south and west) – the two vectors 'span' the whole plane. There is nothing special about north and east vectors. Any two vectors from 0 will work, provided they are not in the same direction, as that would leave us trapped on a single line.

Let's return to my neighbourhood, where my friend Alex lives a few streets away, and the line that connects my house to his is distinct from the line that connects my house to Josh's. I could, if circumstances demanded it, describe any position in the neighbourhood in terms of those two vectors. If a caffeine-starved wanderer knocked on my door asking for directions to a coffee shop, I could point in the direction of Josh with one hand and in the direction of Alex with the other, instructing them to walk a certain distance in the first direction and then turn and head in the second direction until they

reach their destination. The method will always work*
– it's a bespoke satnav, of sorts.

In a flat plane, we cannot do without a second vector,
but we definitely do not need a third. Any proposed third
vector can be reduced to some combination of the first
two and is redundant – its inclusion does not lead us to
any points we could not otherwise reach. So we need our
vectors to be 'linearly independent', meaning that none of
them can be expressed as some combination of the others.

We now have all the ingredients to define a dimension
for any space – it's simply the *least number of vectors
needed to span the space*. The requirement of 'least'
ensures that we end up with linearly independent vectors
that cannot be reduced to one another. In other words,
it reins in our impulse of adding factors that are surplus
to requirements.

At this juncture, some mathematical notation will
help. We can describe vectors in terms of coordinates.
The east vector of the flat plane can be written as (1,
0) – a single unit along the first direction of the space.
The north vector is expressible as (0, 1) – a single unit
along the second direction. Any point in the plane can
be expressed as some combination of these two vectors:
the point (5, 4) is 5 lots of the east vector plus 4 lots of

---

* At least as far as the crow flies – I am assuming that the route
would be unimpeded by buildings or other structures.

the north vector. The point (−3, 6) is −3 times the first vector (remember that the negative value corresponds to the opposite direction) plus 6 times the second. Two vectors suffice to span the entire plane, and two are necessary because, for instance, no amount of the vector (1, 0) can take us to points that have a non-zero 'northern' component.

But what about three-dimensional space? This time, of course, we expect to end up with three vectors that, between them, gesture in three directions. Using the same notation, we can express these vectors as:

$$(1, 0, 0)$$
$$(0, 1, 0)$$
$$(0, 0, 1)$$

Once again, every point in this space can be reached as some combination of our vectors. No fewer will do – because no combination of the first two can reach points with a 'height' – but no more are needed. Three vectors for three dimensions.

I'd wager that you can now write down the vectors that fully specify four-dimensional space. There are no surprises here:

$$(1, 0, 0, 0)$$
$$(0, 1, 0, 0)$$
$$(0, 0, 1, 0)$$
$$(0, 0, 0, 1)$$

This approach is wonderfully scalable: you could proceed to define a fifth, sixth or indeed any number of dimensions if you so desired. These higher-dimensional spaces may be impossible to visualise, but they have been defined mathematically and are no less real than the three dimensions we are more accustomed to.

In practice, when people speak of multiple dimensions, they may not be invoking the formalities of vector spaces in this way. The dimensions of our political beliefs, for instance, may not be represented by literal vectors that we can describe using coordinates. But the two essential components of the mathematical definition – linear independence and spanning sets – are wholly applicable. When proposing one, two, three or any number of dimensions to describe a situation, one must demonstrate first that they can be combined to account for the full range of possibilities, and second that they are truly independent of one another.

Multiple intelligences

The mathematical concept of dimensionality encourages us to think beyond conventional frameworks. The left–right spectrum of political belief, and its economic–social biaxial upgrade, can be thought of as low-dimensional projections of the multidimensional space of public opinion. No collection of two or even three attributes could ever span our range of beliefs. One study, which attempts to classify political beliefs based on voting patterns in Germany, suggests that we would need at least four dimensions to fully describe them.[2] The true number may be impossible to determine, but it need not be restricted to the three dimensions our visual senses are attuned to.

Asking *how many* already represents a step change in how we think of many phenomena. Let's turn to the concept of thinking. One of the most fiercely contested ideas in cognitive psychology is that of general intelligence. Proponents argue that intelligence can be thought of as a singular entity, denoted g. Through the blunt instrument of intelligence tests, one can determine a person's IQ, a single number that somehow accounts for their full range of cognitive skills.

The idea of an all-encompassing intelligence, however, is the epitome of low-dimensional thinking. And nor is it without controversy: the history of intelligence testing has an unfortunate yet undeniable tie to eugenics.[3] Looking to validate their claim of Nordic genetic advantage, eugenicists had long strived to measure coveted traits such as intelligence. After years of failed attempts, in the early twentieth century they stumbled across Alfred Binet's intelligence scale – the basis of IQ testing.

Binet's own intentions were benign: the purpose of his test was to identify French children with developmental disabilities, to ensure they received support in school. In fact, Binet rejected the notion of a single linear scale of intelligence, warning against using his test on the wider population.

Those warnings were not heeded, and IQ testing was seized upon as a sorting mechanism to rid society of its supposedly less intelligent underclasses. The wave of immigration in the United States in the late nineteenth/early twentieth centuries motivated eugenicists' 'race-purifying' policies, which included forced sterilisation to protect society against the 'feeble-minded'. In the first half of the twentieth century, tens of thousands of poor people were forced to undergo sterilisation, often without realising what was being done to them, while scores of them were held in institutions to curb the risk of spreading their 'deficient' genes. In Nazi Germany, euthanasia centres claimed the lives of 100,000 patients with mental disabilities between 1939 and 1941 alone.

IQ testing became pervasive across all corners of US society, from army recruitment to courtrooms; latterly, the decision to execute defendants has often hinged on whether they are deemed 'mentally fit'. While this may seem to protect the vulnerable from the harshest sentences, it reflects the eugenicist notion of intelligence as a marker of one's right to live.

These ideas persist to this day; colloquially, a smart person is one with 'high IQ' while a 'low IQ' person is assumed to have some mental deficit. IQ also adopts other guises. In education, for instance, many countries have developed a fixation on standardised exams that purport to measure students' cognitive abilities through the metric of test scores.

Implicit in IQ testing is the suggestion that intelligence is immutable, yet there is compelling evidence to the contrary. The Flynn effect, named after the philosopher James R. Flynn, points to a consistent rise in IQ scores in many parts of the world.[4] Broadly speaking, IQ rose by about three points per decade throughout the twentieth century. The reasons for this rise are hotly debated, but they all suggest that environmental factors – from diet to what is taught in schools – influence IQ scores. As society has adapted to more abstract kinds of knowledge, and trended towards healthier living conditions, IQ has increased. If intelligence really is static and reducible to a single metric, IQ is clearly the wrong measure.

So what is – or rather, what are – the correct measures? Can we conceive of intelligence across several dimensions? That is the basis of the psychologist Howard Gardner's theory of multiple intelligences,[5] which posits that there are eight intelligences: visual-spatial, linguistic, logical-mathematical, bodily-kinesthetic, physical, interpersonal, intrapersonal and naturalistic. For Gardner, IQ tests our linguistic and logical-mathematical abilities but ignores the other six. While Gardner gives lucid descriptive accounts of each intelligence type, he has received pushback for a lack of empirical data to justify his selections. His framework may be directionally true, but is eight dimensions any less crude? Gardner himself has suggested that there may be room for other intelligences and has flirted with adding spiritual intelligence and teaching-pedagogic intelligence. Our modern times may also have given rise to digital intelligence, which is governed by our relationship with computers.

Those computers, of course, are forcing us to reckon with new types of *artificial* intelligence (AI). As debates rumble on regarding the extent to which machines can think, it is widely thought that to whatever extent machines exhibit intelligent behaviour, it is markedly different to human intelligence. While the artificial neural networks underpinning today's AI are loosely modelled on the human brain, the comparisons are not to be taken literally. Humans do not learn by ingesting unfathomable amounts of data, and our cognitive skills do not boil down to simple mechanisms such as predicting the next word in a sequence, which is the basis of the large language models that have taken the world by storm in recent years. Per Gardner's framework, we derive much of our intelligence from participation in rich social and cultural environments, which can hardly be said of artificial intelligences that are governed by optimisation algorithms.[6]

There is no doubt that the capabilities of AI are advancing in multiple directions, and there's an undeniable intelligence lurking under the hood of these systems. In fact, with multimodal chatbots that can handle text, sound and visual inputs simultaneously, it may be more prudent to refer to artificial *intelligences*.

There is evidently far more to intelligence than monolithic terms such as IQ or AI suggest but, to be clear, the remedy is not to replace them with an arbitrary bundle of individual types of intelligence. The solution to low-dimensional thinking is not just to tag on as many variables as one can think of. For one thing, high-dimensional models induce unexpected behaviours. I once heard about a retired footballer (who shall go unnamed) who instructed an architect to double the length of each dimension of a house they were building – length, width and height. The footballer got his wish but was taken aback by the sheer size of his new abode. As he arranged his furniture (which looked pitifully small in comparison to the space), he was left to reflect on what mathematicians sometimes refer to as the 'curse of dimensionality', whereby the behaviours of low-dimensional spaces can no longer be taken for granted. While doubling each length of a house in Flatland would require four times the area, the doubling in our three-dimensional world manifests as an eightfold increase in volume.

One solution to having 'too many' dimensions is a process known as *dimensionality reduction*, which is an attempt to reduce all possible characteristics to a smaller number that are independent of one another, and that avoid the pitfalls of the high-dimensional curse. The algorithms behind the recommendations on Netflix, for instance, could treat every one of its thousands of movies as a separate dimension. But it may make more sense to group movies into genres – Action, Comedy, Romance or any of the twenty-three listed on IMDb[7] – since users' preferences cluster around them anyway. And there may even be other latent connections between genres that reduce the number further still.

Looking through the list of Gardner's intelligences, one cannot help but sense some degree of overlap: logical-mathematical intelligence does not feel so apart from linguistic intelligence, given that maths and languages share many characteristics. And despite the questionable social skills of some former colleagues, I'm convinced that good interpersonal skills only enhance one's aptitude for mathematical thinking. From this perspective, Gardner may be correct in his pluralistic framing of intelligence, but he may just have settled on the wrong categories.

Multiple personalities

Intelligence testing enjoys an unholy alliance with an equally pernicious form of human profiling: personality testing. Most prominent is the Myers–Briggs Type Indicator, taken by some two million people each year,[8] often at the behest of a human resources department on the basis that it promotes self-awareness and growth.

The test was developed during the Second World War by the writer Katherine Briggs and her daughter Isabel Myers, with the aim of assigning women jobs suited to their personalities. The modern form of the test comprises over ninety multiple-choice questions, with each respondent assigned one of two values in each of four categories. One category speaks to how we derive our energy; a person marked as Extroversion (E) does so through social interaction, as opposed to people who prefer solitude, Introversion (I). The other three categories pertain to how we gather information (Sensing (S) or Intuition (N)), how we make decisions (Thinking (T) or Feeling (F)) and our degree of decisiveness (Judging (J) versus Perceiving (P)).

With four binary choices, the test places people into one of sixteen ($2 \times 2 \times 2 \times 2$) categories, using the letters as shorthand. For instance, an INTP can be understood as logically minded, an ESTP as entrepreneurial and an ISFP as adventurous.

Many who have taken the test will die on the hill of Myers–Briggs, swearing by its uncanny ability to capture their personality traits. This is most likely an example of the Barnum effect, whereby we relate vague descriptions to our own individual circumstances. The effect can be seen in astrology, where any one of the twelve star sign readings on a particular day can be contorted to address your life's quandaries.

Despite the widespread adoption of the Myers–Briggs test, it belongs to the same category of pseudoscience as fortune telling. There's scant evidence that the test is predictive of performance at work, and offering a yes/ no option to statements such as 'You tend to sympathise with other people' is clearly lacking in nuance.

Myers and Briggs based their test on the theories of Swiss psychiatrist Carl Jung, but even he had cautioned that his personality 'types' were approximate.[9] His ideas have been misappropriated in much the same way as Binet's aims for intelligence testing; he would surely be unimpressed with how his general descriptions of human psychology have been compressed and commoditised.

From the standpoint of our mathematical thinking tools, Myers–Briggs fails in at least two ways. First, in its reduction of each category to a binary either/or classification. As Jung wrote: 'There is no such thing as a pure extrovert or a pure introvert. Such a man would be in the lunatic asylum.' The data bear this out: if people really could be characterised as either extroverted or introverted, we would expect the test data to exhibit a 'bimodal' distribution, with two peaks corresponding to those two options. Instead what we see is a bell-curve distribution, which points to a much wider range of profiles[10] – perhaps a continuum would be more appropriate.

The second failure lies in its restriction to four dimensions, which returns us to the question of *how many?* In this case, how many traits do justice to the full span of human personalities? One answer is suggested by another popular framework, the Five-Factor Model, which offers openness, conscientiousness, extroversion, agreeableness and neuroticism. Unlike Myers–Briggs, there is strong evidence that this model speaks to things such as career success. Job satisfaction, for instance, has been shown to be negatively correlated with neuroticism but positively correlated with conscientiousness, extroversion and agreeableness.[11] There is also evidence that extroversion is related to salary level and rates of promotion.[12]

That's not to suggest that the book is closed on personality traits – for one thing, this evidence is strongly skewed towards Western contexts, and it may miss out on other key traits altogether. How does this five-dimensional space capture one's *honesty* or degree of *spirituality?* Once again, we have a low-dimensional projection of a space that is more complex than we have so far been able to specify. The question of *how many* encourages us to keep searching for those missing dimensions, and to reflect on whether our chosen ones are truly distinct.

Piecing it together

Can multidimensional thinking go too far? My brother-in-law Haret might suggest so. It is a festive ritual at our house to complete a jigsaw from start to finish. In recent years I've resorted to a multidimensional approach by separating the pieces according to three characteristics: their colour, the number of tabs (pointy bits) and size (small, medium or large). You should be satisfied that my framework is genuinely three-dimensional, in that my attributes are truly independent of one another and every piece can be described in terms of these three features.

To sort pieces into the appropriate categories, I first create a large grid within a jigsaw tray, with each row representing a different colour and each column a number. To introduce the third dimension, I place three separate trays at different heights; the largest pieces go on the highest tray, the medium-sized pieces on the middle tray, with the smallest pieces on the lowest tray. Thus I can easily identify a piece as being, say, 'large, blue, with two tabs' based on which part of which tray it has been taken from.

I once toyed with adding a fourth dimension, dividing each of my piles once more based on whether they contained any text (I'd spotted pieces of all previous descriptions both with and without, so this too was a legitimate dimension). My plan was to use different coloured trays for the piles as a way of marking this additional characteristic (white trays for pieces with text, red trays for pieces without).

That plan was thwarted when Haret, a master of the craft who insists on plucking pieces from large, messy piles (which he does effortlessly). When it comes to jigsaws, ours is an uncomfortable alliance. The contrast in styles makes for little coordination between us as we execute our methods in silence. Tensions bubble over as we each feel undermined by the other: Haret's system is too holistic for my liking, mine is too convoluted for his – hence his refusal to indulge my extra layer of subdivision.

The truth is that my system is a mathematical work-around because I do not have Haret's eye for jigsaws. His approach evidently does not require the full three dimensions of my scheme – let alone a fourth. If my methods share the bluntness and regularity of a jigsaw piece, then Haret's pattern-matching thought processes have a degree of complexity and elegance that is harder to pin down. They may be more akin to fractals, which have dimensions of a type we have yet even to consider, and which we'll meet in Chapter 10.

9

Distance

*Rethinking proximity, and why our life
goals may be closer than we think*

My experience during Covid-19 was not without irony.
The pandemic emerged just as my wife and I were buying
our first house. Our move from Oxford to Didcot was
motivated by its position on the train network. It would
shorten my commute to London by fifteen minutes
(which, multiplied by two times a day and several days a
week, amounted to a significant saving). We moved into
our new home the day before the first national lockdown
was declared, by which time my employer had no choice
but to resort to a fully work-from-home set-up. In an
instant, my daily commute became a non-event – I would
tumble out of bed and straight to my work desk (hasti-
ly positioned in our conservatory). Didcot suddenly felt
like the most arbitrary of locations.

Prior to the pandemic, my office was conveniently situated on Platform 1 of Paddington Station, meaning that I
could avoid the London Underground and while away the
train journey with uninterrupted work or a good book.

When a meeting or event took me into central London, the days – and the commute – felt much longer. The distance may have amounted to just a few extra miles, but underground travel made it feel disproportionately more tedious. The London Underground map itself distorts our perception of distance because it does not reflect actual distances between stops. It is an example of what mathematicians call a *topological map*, with each location represented as a point and the connections between them represented as lines, and no measurements provided.

More than any event, the pandemic has altered our perception of the work commute. When Covid-19 was at its peak, many organisations had no choice but to adapt to a fully remote set-up, with half of all workers in Great Britain able to work remotely.[1] Having colleagues in the same physical space has obvious benefits, though they are only realised when an office culture encourages interaction and rewards collaboration. Those dynamics, many companies realised, can often be replicated – and even enhanced – in a virtual setting. Teams that nail this down can work productively when separated by oceans and time zones; those who don't may find themselves clambering for a common identity even despite occupying a shared physical space.

If remote work has taught us anything, it is that we need a broader concept of proximity that goes beyond its usual literal interpretation. The mathematical notion of distance that we explore in this chapter provides many such alternatives. It is not limited to geographical contexts either – we will see how the idea of distance can be applied to images and text (a central component of large language models, which are at the heart of AI), to relationships, and even to our ideas of social justice.

One obvious distance metric and two more

As always in maths, to make any headway with our concept we need to lay down some ground rules, or axioms, that it should abide by. Distance is a function of pairs of objects and the following four axioms represent a common set of behaviours that we would expect of any function claiming to represent a distance (what we will term a 'distance metric'):

- The distance from any point to itself is zero.
- The distance between any two distinct points is positive.
- The distance from any point A to a second point B is the same as the distance from B to A.
- For any third point C, the distance from A to B is at least the sum of the distance from A to C and the distance from C to B – this so-named *triangle inequality* says that detours never result in a shorter path.

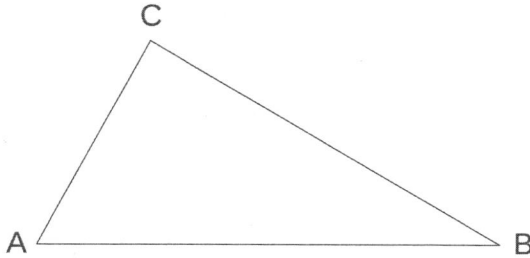

The triangle inequality: the direct path from A to B is shorter than the sum of the paths from A to C and C to B.

Let's start with the most familiar of the distance metrics. Suppose for the moment that we are in a flat plane and define the distance between two points to be the length of the straight line that connects them – 'as the crow flies'. This is usually termed the *Euclidean* distance, after the Greek mathematician Euclid who first documented the rules of plane geometry. But what other version of distance can we cook up?

Towards the end of my weekly five-kilometre Parkrun, the course takes a sharp ninety-degree turn to the right. In usual conditions we just skip diagonally across a patch of grass that the path encloses, but in wet weather the grass is cordoned off and we have no choice but to stay on the footpath and take that sharp turn. Much to the chagrin of runners chasing a personal best, this adds distance to our usual route. The reason is that we're taking a detour to get from one end of the diagonal to the other – the triangle inequality in action.

For a taxi driver operating in the square grid of New York City, the situation is familiar: the only way to get from one place to another is to traverse that grid, going up, down, left or right. When the diagonal is no longer in play, the distance between two points is more appropriately defined as the sum of the vertical and horizontal gaps between them. In homage to the Big Apple, this is sometimes called the *Manhattan distance*, which, unlike the Euclidean distance, welcomes multiple paths. The Euclidean measure is based on the shortest path from A to B: a straight line, with no deviation. But a taxi driver can criss-cross their way along any number of chosen paths, dividing their journey into multiple vertical and horizontal slices (assuming they only move up and to the right). We might ask how many paths are possible for a given map, which borrows ideas from Chapter 7 as it ultimately boils down to the selection of vertical and horizontal chunks. Needless to say, whatever the shape of the chosen journey, the Manhattan distance remains the same; the sum of the vertical slices equates to the 'vertical gap' between the points and does not discriminate between the different routes on offer. Likewise for the horizontal slices. We can show that the Manhattan distance adheres to the four axioms, which makes it a go-to alternative where the most direct route is not available. For instance, for any journey from A to B on the taxi-cab grid, the vertical and horizontal gaps do

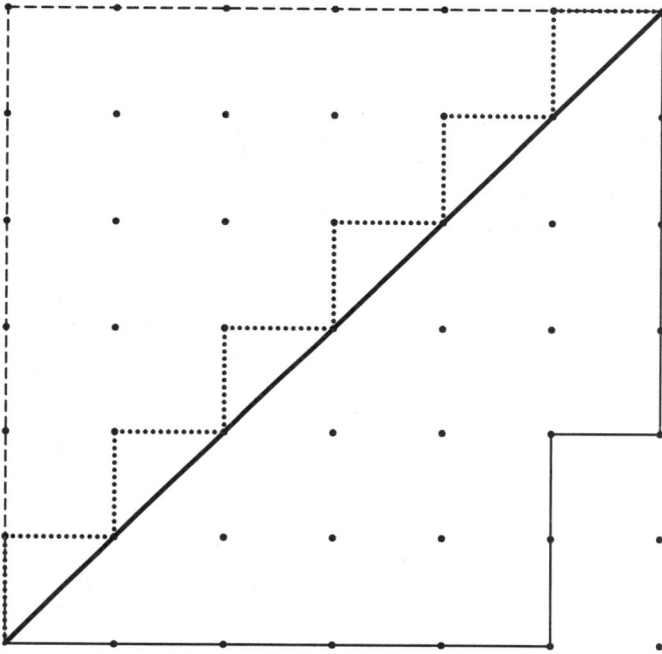

The diagonal line represents the Euclidean distance between the points on the bottom left and top right of the grid. The other lines show three other routes for making the trip without diagonals. The Manhattan distance is just the sum of the vertical and horizontal gaps that separate the two points.

not change when we reverse the journey from B to A, which means that the Manhattan distance is the same in both directions, as per the third axiom. The other three axioms can be checked in much the same way.

The third distance metric we'll explore is known as the Chebyshev distance. Once again, it looks at both the horizontal and vertical gap between two points, but this time the measure is defined as the larger of the two. Once again it meets the four requirements of a distance metric, and once again it gives us an alternative way of thinking about proximity, drawing our attention to the component – vertical or horizontal – where we are furthest from our destination.

An example will help to illustrate the differences between the three distance metrics. Suppose we have two points, which I'll denote with coordinates: the first is the origin (0, 0) and the second is the point (3, 4). The Euclidean distance between the points is the length of the direct, diagonal line that connects them. You may remember from school that another way to compute the Euclidean distance is to apply a formula attributed to the Greek mathematician Pythagoras.[*] The formula relates the length of a line that connects two points to their vertical and horizontal distance. More precisely, the length of this line is given by the square root of the sum of the squares of the vertical and horizontal lengths. The Euclidean distance between our points is therefore $\sqrt{3^2 + 4^2} = 5$. The Manhattan distance is the sum of the vertical and horizontal gaps, which is $3 + 4 = 7$. The Chebyshev distance is the larger of these two gaps: 4.

---

[*] Although the earliest known traces of the result go back to 4,000-year-old Babylonian tablets.

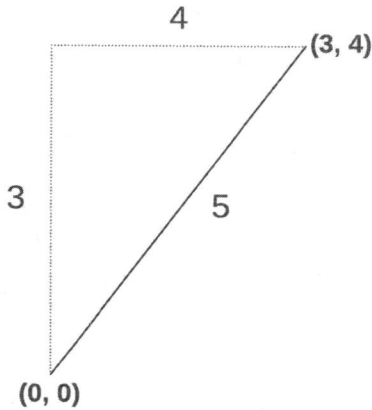

Another way of visualising the differences between the distance metrics is to compare the circles they give rise to. A circle is defined in terms of its centre point and a fixed distance ('radius') from that point to a boundary ('circumference'). The Euclidean distance, being the most conventional, corresponds to the 'round' circle we are familiar with. The Manhattan and Chebyshev distances also give rise to circles, but they are not round. The circle afforded by the Chebyshev distance is the collection of points that are either vertically or horizontally separated from the centre point by a fixed amount – otherwise known as a square! For the Manhattan distance, the circle corresponds to all points where the sum of the horizontal and vertical gaps to the centre point is the same – a diamond.

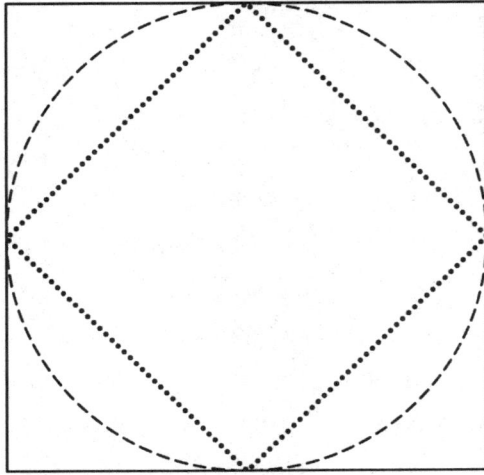

So far, we have restricted ourselves to a two-dimensional framing and looked only at 'vertical' and 'horizontal' gaps. But having laid the groundwork for vector spaces in the previous chapter, we can elevate these ideas into higher-dimensional spaces. The distance axioms remain as they are (in fact, they apply beyond 'vector spaces' that have dimensions attached to them). As always, the pay-off of accessing high-dimensional spaces is that we can apply newly acquired ideas of proximity to a wider range of contexts.

A nice feature of the Pythagorean formula is that it naturally extends to three dimensions. If two points are separated along three coordinates by lengths a, b and c, the distance between them is $\sqrt{a^2 + b^2 + c^2}$. Now let's move into the fourth dimension – and beyond. Taking these formulas as our inspiration, we can now *define* the distance between two points, regardless of how many dimensions are in play. Let's say we have the following two points in four-dimensional space:

$$(1, 2, 3, 4)$$
$$(3, 1, 5, 8)$$

The distances, one coordinate dimension at a time, are 2, 1, 2 and 4. Now execute the formula – that is, square each term, sum them and square root the total: $\sqrt{2^2 + 1^2 + 2^2 + 4^2} = 5$. So there we are: two abstract points, yet separated by a tangible distance.

Extending the other two distance metrics to higher dimensions requires even less work:

- The Manhattan distance between two points is the sum of all the gaps in their coordinates. For our four-dimensional example above, we get a distance of $2 + 1 + 2 + 4 = 9$.
- The Chebyshev distance is the highest gap among all the coordinates. In our example, it is found in that fourth coordinate – a distance of 4.

The differences between the three distance measures reside in how they process the difference in each dimension. Each measure prioritises the dimensions in a different way, and each therefore suggests a distinct way of thinking about proximity. To illustrate the differences in a real-world context, let's take our professional development as an example and imagine a number of different skills, each represented as a separate dimension. Our goal is to become a model employee, which is represented as a distant point relative to our current position. Our aim is to reduce the distance between where we are and this future version of ourselves, and our choice of metric amounts to deciding which skills we should focus on.

The Euclidean distance will advise us to take the most direct route. Reducing the Euclidean distance to a certain destination involves reducing the gap for every coordinate. It ensures that you pay attention to every single dimension at the same time; in relation to our example, it means not neglecting any of the skills that your idealised future self possesses.

The Manhattan distance places equal emphasis on all possible directions. If you reduce the gap by a fixed amount in either the second dimension or the fifty-seventh, you are narrowing the overall gap by that amount either way, even if the gaps in the remaining dimensions remain the same. In relation to our career ambition, this metric adopts the democratic view that every skill is equally worthy of your attention, and that addressing any one of them is adequate, even at the expense of others. Each bit of upskilling will propel you, in equal measure, towards your goal.

The approach advised by the Chebyshev distance is to focus *squarely* on your weakest skills. Afraid of public speaking? That's where you need to improve, even if it means neglecting your development in every other area. The Chebyshev distance puts a premium on the largest discrepancy across all dimensions. When applied to our example, it suggests that if any single dimension sticks out like a sore thumb, you should attack it with gusto. As you develop your speaking skills, another skill will take its place as your weakest – and that's the one to focus on next. To reduce this distance is to confront head on the obstacles that are holding you back most severely. It exposes our largest gaps and shoves us towards a state of balance.

Proximity for text, images ... and people

If you look at two images, you might form an impression of how similar they are, depending on what features your eyes are trained on. You might inspect the images at a high level, with a sense of their overall structure, or you may be drawn to small differences. You can do the same with two passages of text, although it is more difficult to form at-a-glance impressions because text is not visual. But these limits of perception can be overcome by our mathematical notion of distance. Both images and text can be described as points in high-dimensional space, which means we can apply our distance measures and quantify how similar two images, or two passages of text, are to one another.

If the image is, say, an array of 10 × 10 pixels, then we can describe it with a 100-dimensional vector, with each dimension corresponding to one of the 100 pixels. The value for each of those 100 dimensions is typically a whole number between 0 and 255 that captures the intensity of that pixel's colour (0 for no brightness, 255 for full brightness).* What does it then mean for two images, represented by vectors in this way, to be in close proximity to one another? As we've already seen, the interpretation depends on our choice of distance metric.

Consider a facial recognition app that compares two pictures of the same person, taken seconds apart. If the person shows a slight smile in the first but a neutral expression in the second, then the app needs some way of recognising that these subtle changes make no difference to the person's identity. The Manhattan distance, which merely sums the positive differences between each corresponding pair of pixels, overlooks these minor variations and is therefore ideally suited to the task.

The Euclidean distance, which pays more attention to individual differences by squaring them, is more sensitive

---

* Each pixel's colour intensity is represented by an eight-bit number, which can store 256 possible values ($2^8 = 256$), from 0 to 255.

to slight variations in lighting and facial expressions. It is more appropriate in big-picture situations, when our goal is to compare the overall structure or content of images (say, classifying a picture as a dog or a cat) rather than being distracted by the small details.

Since the Chebyshev distance is based on the largest pixel-level difference, it is suited to contexts where individual discrepancies need to be flagged up. It is a good choice when checking for product defects – for example, it will highlight an individual faulty pixel on a screen even when the rest of the product is flawless (faults that may be disguised by the aggregate approach taken by the other two distance metrics).

Let's now move to text, where it is individual words that are treated as vectors,* and each dimension represents some semantic property of the word, such as its grammatical structure or its meaning. The underlying mechanism behind large language models amounts to embedding these words in a high-dimensional space in such a way that each of them sits in close proximity to other relevant words. The tricky part is taking into account different meanings of the same word. In the two sentences 'The book sits idly on my shelf' and 'Make sure you book us a table for two', the word 'book' respectively

---

* Strictly speaking, it is tokens – parts of words – although the distinction matters very little here.

takes on the role of noun and verb. In the first case, we'd expect to see it situated close to words such as 'page', 'blurb', 'library' because these words relate to the meaning of 'book' in that context. When it appears as a verb, we might expect words such as 'restaurant', 'schedule', 'dinner' to appear close by.

This idea is the basis of large language models, the innovation behind generative AI chatbots that respond to prompts by producing whole passages of text. Large language models are able to analyse the interaction of each word in a prompt with the text that surrounds it, extracting the context behind its appearance in a particular sentence. 'Book' starts out as a generic vector and is then shunted around the high-dimensional space until it settles in a position that reflects its intended meaning.

This only works, however, if we have a meaningful notion of distance between two words (or rather, the two vectors representing them) in the first place. One challenge with word embeddings is that certain elements of the vector may be distorted by factors that have very little to do with the actual meaning of the word. For instance, a word that appears very frequently in a text tends to be moved around a lot as the model is trained and ends up having large values across many dimensions. With respect to the kinds of distance measures we have looked at so far, this means that high-frequency words end up being very far away from other words that have similar meanings but appear less frequently. The word 'automobile', for instance, is not used anywhere near as often as 'car' in text sources, so these two words may end up further apart than their actual meaning would suggest they should be.

Adding to this difficulty, the so-called 'curse of dimensionality' obscures our notions of distance in high-dimensional space. Imagine you are the governor of a one-dimensional community of a four-mile strip. Being the conscientious person that you are, you decide to install a grocery store that is within a mile of any given citizen, regardless of their position on the strip. How many stores do you need?

Two will do it, placed at the first and third miles along the strip (denoted by circles below):

But now suppose your town exists as a taxi-cab-style, 4 mile × 4 mile square. To fulfil the same requirement requires twelve stores, as shown below. Wherever you are in this square, a store can be found within a single unit (mile).

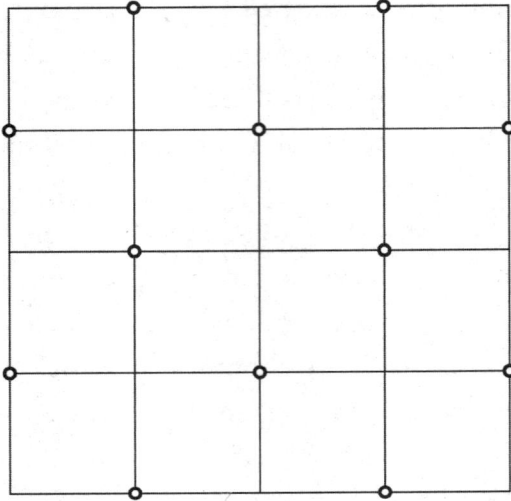

As this simple example hints at, the number of stores rises by orders of magnitude with each dimension; achieving the same degree of concentration requires an exponential increase of points. The upshot is that high-dimensional space tends to be sparse, with points spread far apart. It becomes harder to distinguish between those that are 'close' or 'far apart'; in these spaces, points are always a significant distance away from one another. There are many ways around this issue – one is to reduce the number of dimensions by identifying the salient features of what you're trying to compare; in image processing, for instance, we might extract higher-level features such as edges or shapes rather than analysing objects pixel by pixel.

Another workaround is to opt for a different distance measure altogether, which is based less on the exact position of the vector and more on its direction. Again, it helps to visualise things in two dimensions. Given any two points, we can draw a line from the origin, (0, 0), to each of those points. We can then define a distance measure between these two points in terms of the angle between the two lines. This is known as the *cosine distance*, so called because it depends on the trigonometric cosine function.*

The main point, which holds in any number of dimensions, is that this measure of distance depends solely on the direction of the vectors and not how far away they are from the origin. This makes the cosine distance less susceptible to those issues that arise in high dimensions.

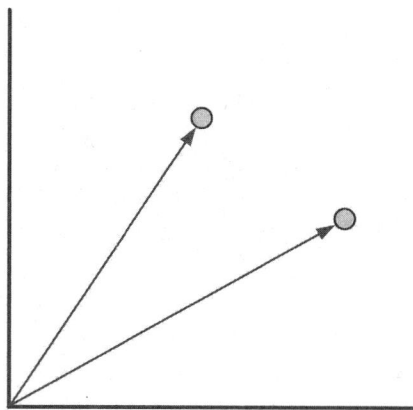

---

* In formal technical speak, one takes the dot product of the two vectors and divides by the product of their lengths, which, in two dimensions, corresponds to the cosine of the angle. The cosine distance is then the result of subtracting the cosine similarity measure from 1.

The cosine distance is not, strictly speaking, a distance measure. Remember that one of the four axioms any distance measure must satisfy is that any two distinct points must have a positive distance with respect to the measure. This doesn't work for the cosine distance: any two points on a line that passes through the origin have a cosine distance of zero (because they have no 'angle' from each other).* But the fact that cosine distance gets around problems such as high-frequency words being very far apart from low-frequency words, as well as the curse of dimensionality, more than makes up for this.

Can we play this game with people – that is, embed ourselves in high-dimensional spaces and apply distance metrics to find out our closeness to one another? We might only have vague descriptions of what those dimensions represent, but that's no different to word embeddings, where the dimensions only partially relate to semantic meaning. So let's run with it: imagine yourself as a vector seeking to establish some measure of closeness with those around you.

The mathematical descriptions of proximity can be applied to the criteria for how we tolerate differences with others. The degree to which you hit it off with someone obviously has a lot to do with how closely

---

* Cosine similarity also violates the triangle inequality.

aligned you are in your beliefs and interests. A tolerant approach to companionship means accepting the odd discrepancy here or there (à la the Manhattan distance, which, you'll recall, simply sums the gap in each dimension and doesn't allow a gap in any one dimension to dominate). Some of us may exhibit a tendency to dwell more on the differences (à la the Euclidean distance, which squares the gap in each dimension and therefore gives more prominence to large gaps, and the Chebyshev distance, which is measured solely in terms of the largest gap in any dimension). Struggling to bridge that gap, we might invest more effort with peers whose outlook on life is more consistently in line with our own.

If we do find ourselves dismissing potential friendships, it may be that we are cursed by dimensionality and judge others according to so many parameters that we perceive our gaps to be larger than they actually are. This may call for an exercise in dimensionality reduction, choosing to ignore certain elements – after all, does it matter if they don't share my enthusiasm for fitness? Or perhaps we can learn to combine these dimensions and gain a better understanding of how others see the world. One's aversion to fitness, for instance, may feel more acceptable when understood in the context of a cultural upbringing that placed little emphasis on physical activity.

Recall how word embeddings are not fixed; they move around as we learn more about how they relate to their surrounding text. Indulging the analogy, we may think of ourselves as floating vectors, moving around as our personalities evolve and new experiences afford us new ways of seeing the world. You may perceive someone as being far away from your present state, but the distance you perceive may be one of magnitude, which may be the wrong thing to focus on. What may ultimately matter more is their direction of travel – they may not share your passion for fitness but so long as they are leaning towards your general direction, those differences ought to be reconcilable.

You could, if you insist, resist any attempts at reconciliation and apply the most egotistical distance metric. I can even offer a mathematical description: just define the distance between two points to be 1 if they are different, and 0 if they are the same. As trivial as the definition may be, it adheres to the four distance axioms. It is called the discrete metric because of the kind of space it gives rise to – not a vector space of continuous points, but a space in which each point is cordoned off from another. You can apply it to any set of objects, but it fails to capture the idea of the continuum, where points can be arbitrarily close to each other. In fact, this metric shows zero tolerance for difference. It is the distance metric of choice for those who treat all differences as equally significant and keep everyone at the same arm's length. The only way to get close to me in this space is to be me. Thank goodness for the alternative distance metrics that foster a more inclusive ethos.

Outliers

Distance functions are a hot property in machine learning, the broad subfield of AI that has given rise to chatbots and agents that can learn across a variety of contexts. Machine learning allows computers to solve problems without having knowledge explicitly programmed into them. The approach rests on feeding systems huge droves of training data and then using fitting curves in high-dimensional data clouds. How this curve fitting is done reveals much about the underlying assumptions driving these models.

Take the simple example of drawing a 'line of best fit' through a scatterplot of points in the plane (the line that most closely approximates the points). Assuming that the points do not all reside on the same line, any line you propose will be subject to some degree of error. Your line is a predictive model; for each point in the scatterplot, it predicts the second coordinate based on the value of the first. The overall amount of error can be thought of as the distance between two vectors: the first vector contains the true values and the second contains the predicted ones.

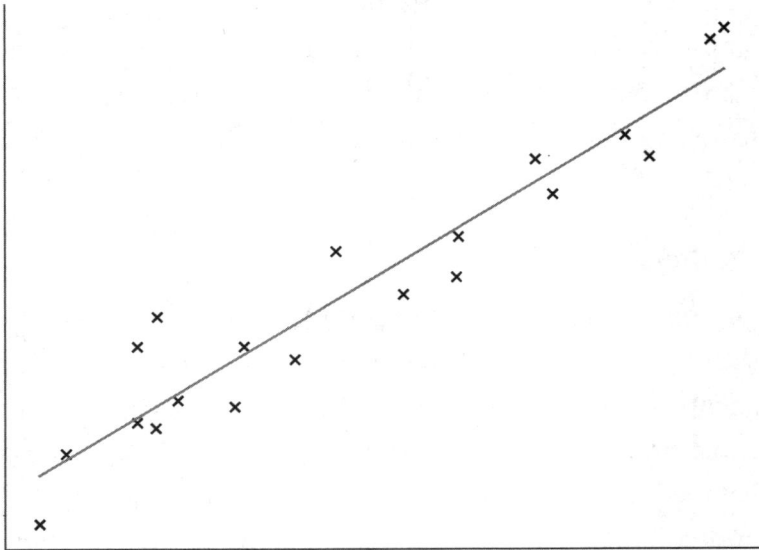

By now, you know that the idea of distance invites a choice between different metrics. For each point in the scatterplot, we first calculate the vertical gap between the actual point and the line. The three distance measures we have looked at each give us a way to calculate the overall error of the model:

- Manhattan distance: sum these values.*
- Euclidean distance: sum the square of these values, then take the square root.

---

* Strictly speaking it is the absolute values (with every difference treated as positive) that are summed.

- Chebyshev distance: take the largest of all these values.

The same approach applies to more complex models. Each distance metric gives rise to a distinct 'loss function' – a mathematical description of the error. In Chapter 2 we met the gradient descent algorithm, which belongs to a family of algorithms for minimising the loss (or error). But this very notion of 'minimising' depends very much on which distance measure we have selected.

Our concern here relates to what the different distance metrics say about our attitude towards outliers, the points in the training data that deviate from the rest. The question is whether they should be ignored, so as not to allow a handful of exceptions to distort the model, or whether they should be noticed, so that the model is better equipped to deal with such abnormalities.

Each distance metric employs a different tack with outliers. Since the Manhattan distance simply adds all the individual errors, an outlier here or there will have limited influence on the final model. The Euclidean distance, which squares each vertical gap, is more easily swayed by such aberrations, and the final model will, in effect, try to get its curve closer to these points to reduce the impact of the errors they give rise to. If the model is ever applied to a similar data point in the future, it will be better placed to make an accurate prediction. With the Chebyshev distance, the model will look to minimise its worst error, lurching towards the most exceptional data points in such a way that they no longer appear as extreme.

The extent to which a model is influenced by outliers reflects a value judgement. If a model is predicting something as asinine as whether an image is a dog or a cat, we may not care very much – and may even find amusement in its tendency to misclassify a cat with canine features (or vice versa). Let's raise the stakes and suppose the object of prediction is the likelihood of a convicted criminal to reoffend, and the output of the model has a major influence on whether to send the offender back to jail, and the length of their sentence. In this case, outliers may represent those who buck the trend in the most unexpectedly positive way. These people may share many characteristics with repeat offenders yet somehow defy the aggregate tendency of this group to reoffend. The question we'll be faced with is what predictive judgement to make for a future offender who displays many of the same characteristics. Do we judge them by the overall tendency, thereby overlooking that outlier's experience, or do we take inspiration from it and adopt a more optimistic view that our offender can be similarly rehabilitated?

Machine learning has been criticised – and rightly so – for its tendency to produce self-reinforcing feedback loops.[2] If a predictive policing algorithm identifies an area as a hotspot of criminal activity, it may prompt higher levels of surveillance and more aggressive policing, resulting in more arrests. A mindless machine algorithm will treat this pattern as vindication of previous judgements, rather than a natural consequence of disproportionate policing. It makes no effort to understand the root cause of human behaviour, such as chronic underinvestment in communities or high levels of poverty. The individuals who reject the worst behaviours expected of them by algorithms are most worthy of our attention because they remind us that criminality is not an inherent trait of any particular group of people, and that our behaviours can be dramatically altered through a change of circumstance.

The technical subtleties of distance metrics translate into our judgements of real people, and to the degree of charity we are willing to show them. Proximity is thus inexplicably tied to our notions of fairness and justice. This connection is made explicit by lawyer and activist Bryan Stevenson who, reflecting on his defence of inmates on death row, writes:

> Proximity to the condemned and incarcerated made the question of each person's humanity more urgent and meaningful, including my own. I went back to law school with an intense desire to understand the laws

and doctrines that sanctioned the death penalty and extreme punishments ... it all became relevant and important.[3]

This is a stirring call for empathy. When people exist on the margins of society, the reality of their suffering is easily diminished in the minds of the public. This disconnect takes on dark forms when marginalised communities are excluded from the scope of justice. Genocidal sentiments are an extreme, yet perfectly natural, consequence of our psychological distance from their suffering.[4]

Is there such a thing as 'over-proximity', where attachment to a particular group identity impairs our moral judgement? There is certainly evidence that believing your group is superior to others makes you more likely to violate your moral ethics. For instance, if people in that group are accused of a crime, then since this threatens the group's perceived superiority, other members are more likely to resist calls for justice. In one online experiment, participants read a newspaper article depicting four alleged cases of military personnel belonging to US-led Coalition forces torturing and killing Iraqi civilians in a prison near Baghdad. When the perpetrators were described as US soldiers, subjects who agreed with glorification sentiments such as 'the US is better than other nations in all respects' were more likely to minimise the emotional toll on victims and dehumanise them.[5] This is not to suggest we discard our group identities, merely that we should be wary in declaring superiority over others and rein in sentiments driven by rhetoric.

Proximity with outliers was the guiding principle behind one of my more interesting choices in graduate school. Two high-profile courses were on offer one semester. The first was organised in partnership with *Sesame Street* and looked at the design principles underpinning the educational content of the programme. The second was a crash course in consultancy frameworks, delivered by the chief education advisor of Pearson. It was heart versus head: an enthralling deep dive into a television show that shaped my early childhood, or a chance to learn the tools of the consultancy trade that would brighten up any CV.

In the end, I went with my heart. Over the next three weeks, storytellers, teachers, pedagogues and designers explained how the secret sauce of *Sesame Street* is its ability to get close to its users. Children are the heartbeat of everything they do; no idea can pass through their filter without validation from their target audience. It is in stark contrast to the archetypal management consultant who occupies a safe distance from the frontlines, working out of hotels but never stepping foot in the communities they purport to serve. In my experience working in the education space, this is the distinguishing

factor between high-impact companies and the pretend-
ers – the willingness to roll up your sleeves and get close
to your users.

Outliers remind us of our individuality, and that the
exceptions to every tendency have the most to teach us.
We should be wary of models that gloss over the most
egregious errors; we should demand to know how they
take account of those individuals at the margins – and
how they deliver justice to them.

10

Fractals

The strange geometry that shapes our world

A lot of the maths in this book is *nice*. The graphs are smooth, the shapes regular. They have their roots in the classical geometry of Euclid from ancient Greece. His thirteen-volume work *Elements* that pinned down theorems concerning lines, angles, circles, cones, spheres, areas and volumes, and much else besides, has endured for more than two thousand years, inspiring not only mathematicians but artists, astronomers, philosophers and writers.

But to think of the world as *nice* is to oversimplify it. As the mathematician Benoit Mandelbrot, who features prominently in this chapter, put it a few decades ago: 'Clouds are not spheres, mountains are not cones, coastlines are not circles and bark is not smooth, nor does lightning travel in a straight line.'[1] The true shape of the world is less clear than Euclid's geometry allows. The first hints of this came in the nineteenth century, when mathematicians developed geometries that departed from Euclid's own – forms where, for instance, angles in a triangle can exceed 180 degrees.

The geometry of *fractals*, our focus in this chapter, had its heyday a few decades ago, due in no small part to the arrival of computers that gave us a way of visualising these most irregular of shapes. Fractals evoked a beauty and terror that mathematicians had yet to reckon with. Most unnerving of all was the realisation that these shapes were as representative of our reality as the conventional objects of Euclid's world.

There is little consensus among mathematicians as to what a fractal actually is. We won't attempt to settle the debate here; instead we'll examine how some of their key characteristics can serve as mental models. We will see, for example, how the property of self-similarity – when the same shape reappears at smaller and smaller scales – has inspired modern types of organisational structures. We will pay a visit to *Jurassic Park*, a memorable depiction of a complex system that arises from the simplest rules. We will explore the link between fractals and chaos, where very small actions can have dramatic consequences. What is true for the clouds, coastlines and rivers is also true of the trajectory of our lives.

We begin, though, with the name itself.

Fractional dimensions

The term *fractal* was coined by Mandelbrot in the 1970s, at a time when physicists realised that the usual shapes of geometry were too simple to describe natural phenomena such as cloud formations, mountain ranges, river systems and the distribution of galaxies. A made-up word, it had connotations of fractures and fractions, two defining features of the newly discovered geometry. The *fractures* of cloud formations are evident, but what do they have to do with numerators and denominators? It is all in the dimension.

A point, we know, has zero dimensions. A line is one-dimensional, a square is two-dimensional and a cube is three-dimensional. Can we conceive of something in between these whole-number dimensions?

With fractals we can. We'll start with an equilateral triangle and join the midpoints of its three sides to form three smaller triangles (and one upside-down triangle). Repeat this process for each of the three upside-down triangles that we've just created and continue this way. Here are the first few iterations of this process:

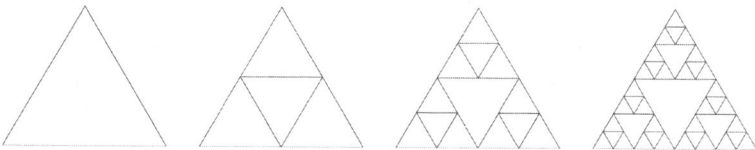

If we allow this process to continue ad infinitum, the resulting shape is the Sierpinski triangle. It looks, roughly, like this (a drawing, however finely drawn, can only ever be an approximation of the infinite iterations it takes to generate the real thing):

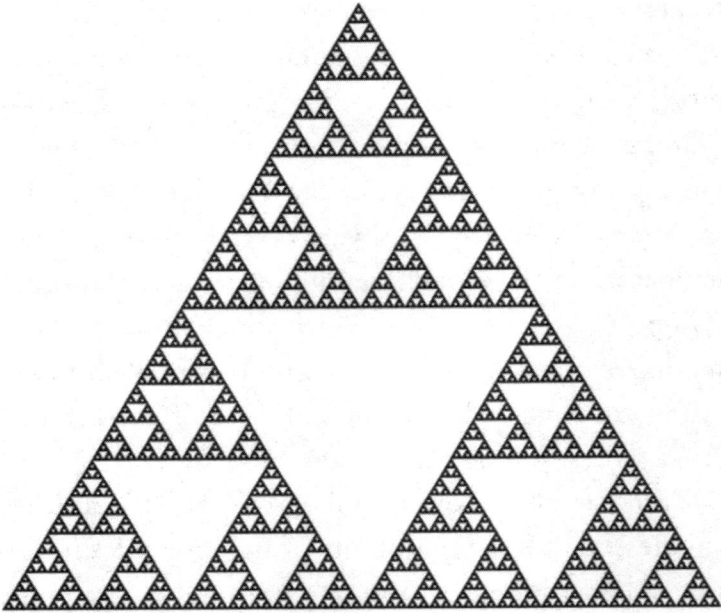

The Eiffel Tower, a lattice structure composed of colossal trusses that branch into smaller versions of themselves, is a three-dimensional approximation of the Sierpinski triangle. The design makes ingenious use of the fact that to deform the structure, one must deform at least one of these sub-members, giving strength and stability with significantly reduced weight.

Let's turn to the dimension of the Sierpinski triangle. Our first thought might be that it is two-dimensional, like a normal triangle. However, through the iterative process the Sierpsinki triangle gives up a lot of area – it is not as 'hungry' for space as ordinary two-dimensional shapes. With that in mind, we might expect its dimension to be slightly less than 2.

Let's make this more precise, by first thinking about how dimensions behave:

- In one-dimensional space, doubling the size of a line increases its length by a factor of 2.
- In two-dimensional space, doubling the side lengths of, say, a square or a triangle increases the area by a factor of 4, or 2 × 2.
- In three dimensions, doubling the edge lengths of, say, a cube (or tetrahedron if you want your faces to be made of triangles) increases the volume by a factor of 2 × 2 × 2 = 8.

Here's the upshot: scaling a shape by a factor of 2 (or any factor, for that matter) has the effect of increasing the size of the new shape by that same factor *for each of the shape's dimensions.*

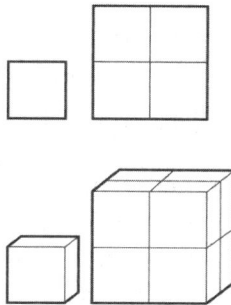

But what happens to the Sierpinski triangle when we double the length of its sides? Below we have a Sierpinski triangle alongside a second, enlarged version where the side lengths of the original have been doubled.

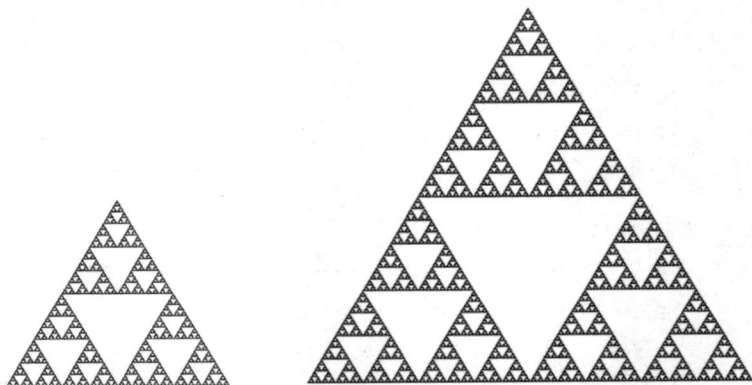

If the Sierpinski triangle really were two-dimensional, we would expect the area of the enlarged version to be four times as large as the original copy (just as with an ordinary triangle). Yet that's not what happens – you may have noticed that the enlarged version can be formed by joining together three exact copies of the original triangle (at the top, lower left and lower right). The enlarged shape therefore has *three times* the area of the original Sierpinski triangle.

To summarise: the Sierpinski triangle is not one-dimensional because doubling its side length more than doubles its area, but nor can it be two-dimensional because its area does not quadruple. We are now ready to calculate the exact dimension: we need to how many doublings result in that enlargement factor of 3. In terms of an equation, we are looking for the value d for which

$$2^d = 3$$

If you've studied logarithms, you will recognise the answer as $\log_2(3)$; if that means nothing to you, then you simply need to know that this value corresponds to approximately 1.59. So that is the dimension of the Sierpinski triangle – not a whole number but an unwieldy one that sits somewhere between 1 and 2.

The Sierpinski triangle illustrates the first defining property of a fractal – a dimension other than a whole number.* This number is a measure of a shape's irregularity; it signals the degree to which the shape fills the space available to it. Where a square or ordinary triangle grabs every available point, the Sierpinski triangle casts many of them aside.

Fractional dimensions are not just a mathematical oddity. In the 1950s, the mathematician and physicist Lewis Richardson was exploring the intriguing question of how the length of a border influences the likelihood of

---

* Strictly speaking, $\log_2(3)$ is irrational – that is, it cannot be expressed as a fraction. The term 'fractional' is being applied loosely to denote the fact that the dimension is not a whole number.

two neighbours going to war. During his investigation he noticed something curious: the recorded measurements of various borders were riddled with inconsistencies. What he realised, and what Mandelbrot later fleshed out,[2] was that the measurement depends on how much you've 'zoomed in'.

If you took it upon yourself to measure the length of Britain's coastline, you'd have to walk along its beaches, summing each distance your ruler is able to measure. If you had the patience to repeat the exercise with a smaller ruler, you'd have to take more measurements, and your estimate of the total length would increase. Smaller and smaller measurements would reveal more and more detail – bays within bays, peninsulas within peninsulas – resulting in larger and larger measurements, which might suggest that the coastline is infinite!

One way to resolve this apparent paradox is to overlay a grid of square cells on the coastline and count how many cells it intersects with. We can then play the same scaling game as before: when we enlarge the coastline by a factor of 2, we'll find that it now has *more than* twice the number of cells. This tells us that the coastline has more than one dimension: it turns out to be approximately 1.21. There is nothing special about Britain here, of course: all coastlines exhibit this behaviour, and a higher dimension corresponds to a greater degree of 'roughness'. The Norwegian coastline, with its numerous fjords and islands, is particularly irregular, its fractal dimension coming in at close to 1.5.

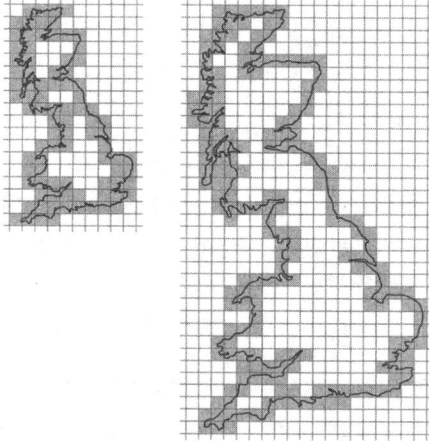

In Chapter 8 we explored the trappings of low-dimensional thinking and suggested that phenomena such as our political beliefs or intelligence may be situated in higher-dimensional spaces. We were still restricted to integers – the *four* dimensions of political thinking, the Big *Five* personality traits, Gardner's *eight* intelligences. Fractional dimensions expand our options further, allowing space in between these clear-cut descriptions. The next time you're walking along the beach, remind yourself that some things are so irregular that they cannot be described by a discrete set of labels. If it is true of coastlines, it is certainly true of the social phenomena that pervade our everyday lives.

Self-similarity

In a very literal sense, fractals are the shapes that keep giving. We can see, for instance, that the Sierpinski triangle is composed of three smaller Sierpinski triangles. If you zoom into any one of those, you'll find three yet smaller copies of the Sierpinski triangle. Because the fractal is generated via an unending iterative process, you can zoom as much as you like, with no blur and no loss of detail – what surfaces at each level of magnification is an exact replica of the same triangle.

This is what it means for a fractal to be *self-similar*. In classical geometry, two objects are said to be *similar* if they have the same shape. You may need to rotate, reflect or enlarge one to get the other, but they basically look the same. Fractals turn that definition inwards; they contain smaller copies of themselves. In nature, self-similarity does not

manifest for infinitesimally small scales, but, for instance, the buds of a romanesco broccoli comprise a whole shape that resembles each individual bud, and each pinna of a fern is a miniaturised, near-perfect copy of the whole leaf.

Self-similarity can also be applied to how we set and pursue our goals. The challenge with medium- to long-term goals is that distant prospects do not always translate into immediate or sustained action. How many New Year resolutions die an all-too-predictable death before January has passed?

Once we break our goals into smaller ones, however, pursued over shorter time periods, then they both seem more manageable and come to shape our everyday habits. I may have my sights set on completing a half-marathon, for instance, but the weekly five-kilometre Parkrun anchors me to a running regimen that accumulates over the span of several months into preparation for the larger distance. From diet to conditioning, my preparation each week is a ritualised microcosm of the real thing. By situating our goals within shorter timeframes, we replace projects with processes. Aristotle was onto something when he quipped: *we are what we repeatedly do*.

For organisations, the geometry of self-similarity shows us how it is possible to pack limitless complexity within a finite amount of space. One of the earliest studied fractals is the Koch snowflake. This time, start with

Think Like a Mathematician

an equilateral triangle and, for each of its three sides, proceed as follows:

- Divide the line into three equal segments.
- Draw an equilateral triangle, taking the segment as the base.
- Remove that middle segment.

As is customary for fractals, repeat to infinity, applying these steps to every line each time. Here are the first four iterations:

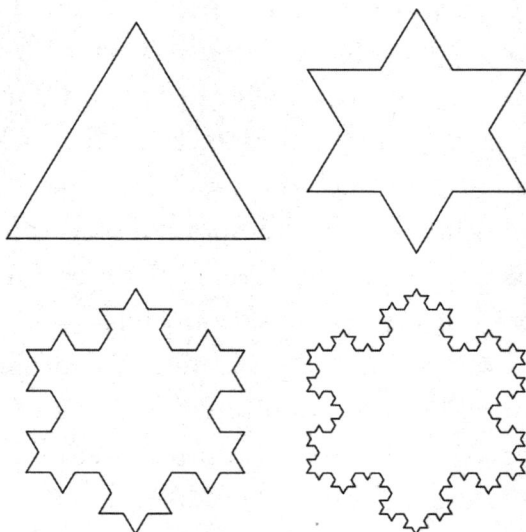

The Koch snowflake is the resulting shape – and the reason for its name is apparent by the fourth iteration. The shape contains a finite area; it can be fully contained within a circle that passes through the three corners of the original triangle. But now consider its perimeter. At each iteration, every line is replaced with four shorter lines, each of which has a length a third of the one it is replacing; collectively, the shorter lines have a total length that is four thirds of the previous line. This pattern will repeat for every line, which means that the perimeter of the shape will increase by the factor of $\frac{4}{3}$ at each step. We have an exponential sequence with a multiplicative factor in excess of 1; thus the perimeter will shoot up to ever-increasing values, surpassing any fixed amount. There is only one conclusion, which you will find nowhere in conventional geometry: the Koch snowflake, while having a finite area, has an *infinite* perimeter (much like the coastline of Britain, except this time we have a purely mathematical object that can be measured with absolute precision). This is the mathematics of self-similarity – strange, twisted, irrefutable.

For my colleague, the maths YouTuber Ayliean, this blend of the finite and the infinite informs a philosophy to live by. Ayliean, who bears a Koch snowflake among many mathematical tattoos, tells me that it serves as a constant reminder that nature is replete with intricately detailed patterns that gesture to the infinite and can fill our lives with meaning. 'Although our lives are short they are infinitely detailed, and there is joy in attending to those details', says Ayliean.

There are growing murmurs that organisations can take inspiration from these properties of fractals when they revamp their traditional hierarchical structures. A 2021 report by Boston Consulting Group suggests that our standard notions of scale are now defunct.[3] In the past, the success of a company was a function of its size: larger companies, with more employees and higher turnover, were in the ascendancy. To scale up, they would hire more workers, build more factories and open new offices, extending into new geographies according to the same top-down operating procedures. According to BCG, this method of scaling is in decline, due to factors such as fragmented geopolitics, digitisation and technological innovation.

If size doesn't matter as much as it used to, then what does? Locality. BCG advises companies to gear up for 'winning multiple local battles instead of waging a single global war'. A company's product and service line must be every bit as diverse as its markets; each battle is an effort to win over a particular customer base. Strategy is dictated not through centralised operating manuals but small-level interactions in each market. Activity takes on a jagged form as the company nimbly responds to each market's needs, a modern way of working known as the *fractal advantage*. In the Koch snowflake, all the action takes place on the boundary, and it is there that its self-similarity arises, giving rise to a limitless intricacy.

A fractal is a perfect illustration of the 'think global, act local' edict that modern companies strive for, generating complex behaviours at a global level through the repeated application of simple rules at a local level. A company is no more, and no less, than the accumulation of these local actions.

This 'modern' approach to organisational structure may not be as new BCG imagined. If we look hard enough, we can see self-similarity in the design of cities. A large city that has absorbed many small villages takes on the appearance of a fractal; the city contains villages, which contain small neighbourhoods and so on. One example is Rome, which exists in stark contrast to the zonal confines of urban behemoths such as New York. A 'fractal city' is a tightly integrated network with roads that connect key central hubs, meaning that daily existence feels more connected and organic. Such cities do not suddenly spring up; they develop over the centuries, through the millions of choices that its citizens make on how to live.

The shape of a city is determined by its rules of governance. Fractal cities are often rooted in 'proscriptive' rules that give people more autonomy (settling parking disputes, for instance, or choosing neighbourhood speed limits), allowing them to live as freely as possible within the laws of the land. This stands in opposition to top-down 'prescriptive' rules. Consider zoning rules, which may designate part of a city 'industrial' or 'residential', limiting activities in areas that fit those labels, resulting in a more regimented structure across the city. Modern transport has made such an approach more viable, as citizens can be expected to compartmentalise their lives in different parts of a city.

But there are trade-offs, both environmental and personal, brought about by the daily bustle of city life. As we look to adopt more sustainable living habits, a retreat to the fractal city may be called for.[4] Its promise is one of increased autonomy for its citizens, and of a richer diversity of culture and experience. In a resource-scarce world, it may be the only design that makes sense.

Complexity from simple rules

The paradox of fractals is that their complexity arises from the simplest of rules. Generating the first iteration of the Sierpinski triangle or the first Koch snowflake is child's play; it is astonishing to think that there's nothing more to their fully realised, self-replicating form than the recursive application of this single manoeuvre at ever-finer scales.

But even these fractals are simple compared to others out there. As disconcerting as it is to imagine the spiky Koch snowflake boundary re-emerging at each scale, it does at least possess a degree of predictability. However much we zoom in, the exact same form reappears.

The same cannot be said for the Mandelbrot set, which has done more than any other fractal to put these shapes on the map. At its 1980s zenith, the bug-like silhouette was emblazoned on dorm room walls, T-shirts and even album covers, drawing people young and old, artistic and scientific, into its orbit.

The definition of the Mandelbrot set is more technical than the others we've seen so far, involving complex numbers – a two-dimensional extension of ordinary, 'real' numbers that allows for the square root of negative numbers. But the recipe for generating the fractal is once again straightforward. We fix a number in this space of complex numbers – let's call it c. Then, starting with the value $z = 0$, we ask what happens when we apply the following operation repeatedly:

$$z \to z^2 + c$$

The trickiness here is in knowing how to multiply and add complex numbers. The first step takes the number 0 and applies the above computation to it, giving $0^2 + c$, which is just c. At the second step, we plug in this new value c and end up at $c^2 + c$. Then we plug in this new value $c^2 + c$, feed the result back in and keep going.

One of two things will happen: either these values will settle within a certain range, or they will fly off towards infinity. Small values of c result in the former and large values of c the latter, but in between the two extremes emerges a stunning picture of complexity, made literal by the Mandelbrot set's colouring scheme: the points c for which a settling occurs are black, while all others are white.* There's a fair amount to absorb here, but at root is that simple recursion: $z \rightarrow z^2 + c$. The resulting beetle-shaped figure is believed by many to be the most complex mathematical object ever conceived.

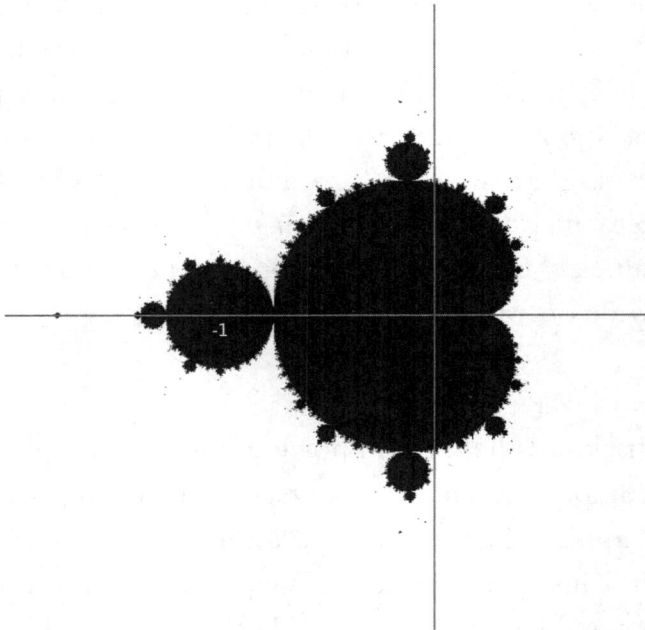

A passing glance at an image of the Mandelbrot set tells you that, as with the Koch snowflake, all the action

---

* In a more intricate version, these points are coloured according to how quickly the sequence shoots off to infinity.

is at the edges. When you zoom in to the spiky forms around the boundary, a kaleidoscopic galaxy of patterns emerges. Unlike the fractals we've encountered so far, the Mandelbrot set does not merely repeat the same structure at different scales. There is a degree of self-similarity – you'll find miniaturised beetle-shaped figures at every scale – but there are also new arrangements. The Mandelbrot is infinite in more ways than one: it not only keeps going, but it keeps changing.

Mandelbrot was not the first person to study the object he is renowned for, but he deserves credit for firing up the public's imagination with the first high-resolution depictions of the shape. Mandelbrot was aided by his association with IBM; his access to computing resources meant that, for the first time, it was possible to visualise recursively generated patterns beyond the first few iterations. When Mandelbrot received his first printouts in 1980, he found that the edges of the shape were smooth; it turned out the lab technicians had taken it upon themselves to clean up the 'fuzziness', assuming it was due to an accumulation of dust particles. Mandelbrot knew better, having anticipated the roughest of edges.

The behaviour at the boundary is highly erratic; if one point lights up, there's no knowing what its closest neighbours will do. There are two seemingly opposed ideas in play: the edge of the Mandelbrot set is purely deterministic – governed by a relatively simple operation ($z \rightarrow z^2 + c$) – yet its behaviours defy attempts at prediction.[5]

Such a notion flew in the face of conventional wisdom – for real-world situations at least, the traditional view was that anything so complex would have to arise from complex causes. A shape as irregular as the Mandelbrot would surely have some form of complexity built into its core. But in the nascent field of *chaos theory*, a stranger reality was taking form; far from being a quirk of abstract shapes such as the Mandelbrot set, chaotic behaviours were being observed across a range of phenomena. Nature seems to make good use of the advantages conferred by fractals; plants make use of wiggly surfaces, for instance, to increase their capacity to cool and absorb nutrients. Mandelbrot himself had glimpsed fractals while conducting a study on the relationship between cotton prices and river floods, with irregular patterns replicating at smaller scales (daily price changes mimicking monthly ones, for instance) – both utterly unpredictable and entirely determined by simple equations. Being British, I am well accustomed to the unpredictability of weather. It would be a mistake, however, to attribute today's rainfall to blind chance. There's no randomness to weather. It is another example of a chaotic system – describable by a handful of relatively simple equations, but at the mercy of unwieldy patterns.

Sensitivity to initial conditions

The difficulty in predicting what happens in chaotic systems comes from how minute differences in inputs can become dramatic changes to the output, in what's known as *sensitivity to initial conditions*. This is not an issue in classical Newtonian physics, where the regular movement of objects – planetary orbits, swinging pendulums, rolling balls – are easily predicted, even allowing for small changes to inputs.

Sensitivity to initial conditions is also known more commonly as the 'butterfly effect', which suggests the extreme possibility that a butterfly flapping its wings in the Amazonian jungle might cause a storm to rage across Europe some weeks later.[6] I'm reminded of the idea every time I play a game of pool. What ostensibly appears to be a classical system – balls whizzing around the table – is more akin to a chaotic one. On rare occasions, my break-off shot is near perfect, yet it is utterly impossible to replicate. One could find equations to model the movement of balls around the table, factoring in things such as their mass or the exact strength with which I strike the cue ball. But the slightest deviation in any of these conditions results in wildly different outcomes.

The idea is a popular theme in literature. Ray Bradbury's 1952 short story 'A Sound of Thunder' speculates on how the single action of a time traveller might alter the course of history in disconcerting ways (for instance, accidentally stomping on a golden butterfly 65 million years ago results in a significant change to language in the present, with people now uttering words phonetically). More famously, chaos theory is the central theme of Michael Crichton's *Jurassic Park*, published in 1990 when the field was entering the mainstream. In the novel, mathematician Ian Malcolm prophetically warns John Hammond, CEO of the bioengineering company InGen, of the dangers of using genetically recreated dinosaurs as the basis for an amusement park. Hammond's hope is that he can tame the beasts and keep them safely locked up. Malcolm, a specialist in chaos theory, argues that this is futile, drawing on the key lessons of his field:

> Chaos theory teaches us … that straight linearity, which we have come to take for granted in everything from physics to fiction, simply does not exist. … Real life isn't a series of interconnected events occurring one after another like beads strung on a necklace. Life is actually a series of encounters in which one event may change those that follow in a wholly unpredictable, even devastating way.

It must be said that the film adaptation does not quite manage this degree of nuance. Viewers also miss out on Crichton's use of the Heighway fractal, a dragon-like curve that gradually morphs from a single line into something far more daunting. Crichton includes successive iterations of the fractal at the start of each chapter, a visual metaphor of how the simplest things can spiral out of control.

The butterfly effect reinforces the counterintuitive idea also present in Tolstoy's reading of history that our lives are determined not by the big but by the vanishingly small (see Chapter 1). When the cognitive psychologist Amos Tversky was asked what got him into the field, his response sounded like that of a chaos theorist. 'It's hard to know how people select a course in life', said Tversky. 'The big choices we make are practically random. The small choices probably tell us more about who we are. Which field we go into may depend on which high school teacher we happen to meet. Who we marry may depend on who happens to be around at the right time of life.'[7]

In fact, chaos theory goes further, suggesting that those small causal factors may be impossible to determine. As I write this chapter, I have a cast on one wrist and a splint on another, having been diagnosed with a scaphoid fracture on both following an eminently avoidable bike accident. The most basic explanation for my fall is that I misjudged a sharp turn at a roadblock and, with one hand gripping a snooker cue case, I was unable to recover my balance with my other hand. There is no excusing my recklessness in cycling without a firm grasp of both handlebars. But what possessed me to carry the cue in the first place? It was a weekday, so my work schedule must have allowed a ninety-minute slot – a cancelled meeting, perhaps. Or maybe my poor snooker form in a previous session left me with an itch to get back on the practice table. The car must have been unavailable because otherwise I would have combined the trip with a visit to the swimming pool. The most chaotic system of all – the weather – was a key determinant: if it had rained that day, I may have gone for a run instead. And if the roadworks had started a day later then, well …

The central lesson of chaos theory is not that small actions escalate into larger ones – after all, not every action results in a broken wrist or a dinosaur stampede – but that our actions exist alongside many others in a system so complex that the effects of individual actions, small or large, are impossible to determine. On some level, we all have to plead ignorance to the true causes of everyday events.

We make sense of the good and bad in life by assigning particular causes and ignoring the hidden minutiae of everyday events. Chaos theory counsels us to think twice before assigning blame, or credit, for a given situation. It also cautions us against judging others. Even if our complex personalities can be reduced to simple governing principles, there's no real knowing what 'initial conditions' led us to be how we are. This is not to let humans off the hook for bad behaviour, but simply to recognise the challenge inherent in empathy. If we wish to walk in another person's shoes, we had better be prepared to venture deep into their history in our search for root causes. Even then, we must accept that the answers may be elusive. Our lives unfold in ways we cannot fully comprehend. The only thing we can be sure of, with mathematical certainty, is that the triumphs we celebrate may have tilted towards tragedy but for the proverbial flapping of a butterfly's wing.

Chaos theory has never felt more relevant to our interpretation of everyday events. As the world becomes ever more interconnected, it also becomes more sensitive to our individual deeds. As the political scientist Brian Klaas puts it: 'We control nothing, but influence everything.'[8] This strikes me as a hopeful realisation; we all impact the world in ways we probably fail to appreciate. It is also a daunting one because with such power comes responsibility. The words we utter, the actions we take – even the seemingly mundane ones like the way we engage with social media posts we disagree with – have the potential to morph into events that lie beyond the event horizon of predictability, affecting others to a degree we are unable to fathom. We should exercise them with care.

Everywhere continuous, nowhere differentiable (aka hidden monsters)

We humans took our time to uncover fractals, but once we had there was no escaping them. The branching network of our blood vessels, shifting wildlife populations, star constellations, stock price data, cloud formations, sea currents ... when one peers closely enough, fractal-like structures can be seen across nature.

That we missed them for so long is hardly surprising, for our minds are attuned to smooth patterns. As Ian Malcolm reminds us in *Jurassic Park*, we incline towards linearity, thinking of each event bleeding into the next. The most terrifying aspect of *Jurassic Park* – more so even than the charging Tyrannosaurus rex – is the realisation that our lives unfold at every scale as a cascading set of interconnected events whose causes and effects elude us. Reality is far more bizarre than we are able to intuit.

You may feel that this overstates the case for fractals – are they really as pervasive as I am suggesting? Mathematicians glimpsed them as far back as the seventeenth century. The German mathematician Gottfried Leibniz, a pioneer of calculus, briefly studied self-similarity and made reference to 'fractional exponents'. As he acknowledged, the geometry of his day did not equip him to go much further. It would take centuries, and the advent of the computer, to fully unravel these shapes and all the strangeness they inhabit.

Even before then, mathematicians brushed shoulders with fractals from time to time. In 1872, an example was found of a function whose graph was different to anything that had been conceived before; in today's parlance, it would be considered a fractal. To make sense of this type of function, we need to recall two concepts from earlier chapters:

- A function is *continuous* if it has no sudden breaks: you can sketch it without removing your pen from the page (Chapter 1).
- If you draw a continuous function, I'm willing to bet it looks curvy and smooth, like the one below. You can draw tangents at most, if not all, points – your function is said to be *differentiable* at those points; it has a gradient (Chapter 2).

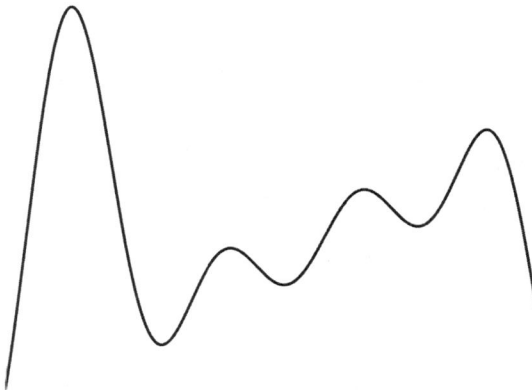

It is possible, however, to have points without a gradient. Consider the lowermost point in the graph below, of a so-called 'sawtooth' function. Looking to the left of that point, the gradient appears to be negative but looking to the right it appears to be positive. Since it can't be both, the gradient at that lowermost point is simply not defined.*

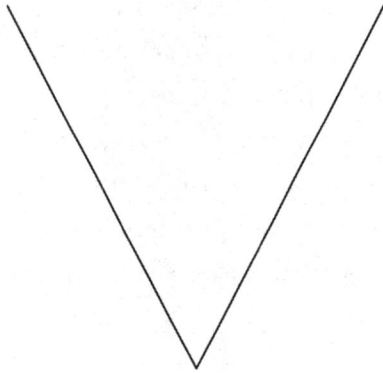

The functions we are concerned with here are known colloquially as *monsters*; their hideousness resides in the fact that they are *continuous everywhere* (no breaks) and *differentiable nowhere* (there is not a single point smooth enough for a gradient to be defined).

---

* This is different to the stationary points of Chapter 2, where the transition from a positive gradient to negative gradient is gradual; here the switch is sudden.

The combination shows a disregard for the smooth-
ness imposed by calculus. It scarcely seems imaginable,
yet we've already encountered the idea; the boundary of
a Koch snowflake has no sudden breaks, but zoom in to
any point and any hope of a tangent fades away.

The idea behind these functions is to turn every point
into something that resembles that non-smooth point of
the sawtooth function. They are actually fairly straightforward to construct – another example of a complex
object arising from a simple rule applied repeatedly.
I have left out the details but here is an approximate
sketch of one such function:

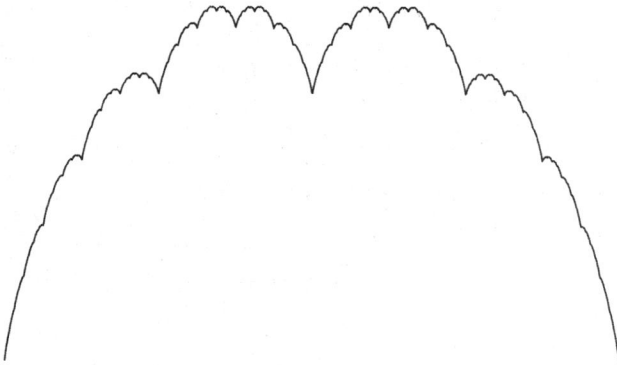

The mere existence of this function unsettles our conventional view of mathematics. The functions we study
in school are nice and smooth. They are the trigonometric 'wave' functions, the exponentials that model growth

and decay. They seem natural because they feature in our descriptions of reality.

But the monster functions are not outliers. Imagine placing the entire collection of continuous functions into a bag and drawing one at random. Since it takes a leap of abstraction to concoct an example like the function above, you would probably expect it to be very differentiable.

Yet it turns out that the vast majority of continuous functions have more in common with the monster function than those intuitive smooth objects. We are in that murky territory of comparing infinite collections of mathematical objects, but it is an indisputable fact that the probability of selecting a function that is nowhere differentiable is 1 – that is to say, a mathematical certainty.

We saw in Chapter 1 that unwieldy non-fractions, or *irrational numbers*, are no mere quirk of the number line. Fractions and whole numbers, though more intuitive, are like specks of dust compared to irrational numbers. The irrationals and non-differentiable functions swarm the mathematical universe like dark matter, concealed from view. They are largely ignored in mainstream education, and some mathematicians will go their whole lives without encountering non-differentiable functions. Is ignorance of them bliss? Fractals have put an end to any such debate by showing us that nature itself is replete with such objects.

Our multisensory experience of the world is just a snapshot of the true nature of things. Our eyes can only process wavelengths between 400 and 700 nanometres, for instance – a tiny sliver of the electromagnetic spectrum. That leave microwaves, X-rays, cosmic waves, radio waves and infrared hidden. Our default view of the world is not only partial; it is also inaccurate. The cognitive psychologist Donald Hoffman has put forward an argument that suggests that our perceptions have evolved to report on fitness, even if that means hiding reality from us.[9] For example, if you see a long, winding object on your lawn you may recoil thinking it is a poisonous snake when it is simply a hosepipe. According to Hoffman's theory, our brains are tuned to make such mistakes because of the existential threats posed by snakebites. Our survival instincts kick in, gladly sacrificing accurate real-world perceptions for the sake of caution.

Hoffman also suggests that maths may be immune to our distorted perceptions of reality. Logic and reasoning offer cast-iron truths that exist independent of our multisensory experience. Yet the monster functions remind us that once we impose a definition in mathematics, we are at the mercy of its logical implications – even those that force us to reckon with the most unnerving of consequences. The smooth mathematical objects that our minds naturally think of no more reflect the subject than our rough-and-ready senses reflect reality. Maybe they come to mind so readily because we've smoothed over the cracks of reality itself, convincing ourselves that the universe is more orderly than it actually is.

If that argument feels too other-worldly, let's come back down to earth – and specifically to social media, where we are constantly applying filters to conceal aspects of our true identity from others. Social media platforms incentivise us to give the most selective accounts of ourselves that will, one way or the other, elicit engagement from others. It is easy to scroll through a feed and think that the rest of society has life figured out – from parents dishing out advice on how to nurture intelligent tots, and productivity gurus intimidating us with their morning routines, to armchair pundits charting singular paths out of geopolitical conflict. The posts deemed most worthy of liking or resharing are also the most declarative, often stating opinion as a matter of objective fact. They leave little space for us to express our vulnerabilities, or our uncertainty about the day's trending topics.

The contours of our daily experiences are far less smooth than our digital projections portray. When it comes to grasping the human condition, interactions at close quarters – whether with friends or families, colleagues or strangers – remain the most effective means of unravelling the messy nuances of daily life. These exchanges may not scale as easily as viral posts, nor will they reveal formulas for success. But they do remind us that very few people have solved life in full, and that for all the success we parade through our online avatars, the vast majority of us are consumed with doubt. The best friendships create a space for us to unearth one another's hidden monsters, however unwieldy or fearsome they may be.

Our exploration of bizarre mathematical objects has taken a philosophical turn as we reflect on the virtues of making the unseen seen and confronting the roughest edges of our identities. In the timeline of mathematics, fractals are among the most modern of objects, unearthed through a combination of human endeavour and computing power. It's all too easy to simply marvel at their beauty and dismiss them as a quirk of mathematics when, in fact, they have been there all along, awaiting our discovery. The more we grasp their unusual forms, the more we learn something of our own chaotic and infinitely complex lives.

Epilogue

Great expectations

An exchange between the mathematician Leonhard Euler and philosopher Denis Diderot is alleged to have taken place at the court of Catherine the Great in the eighteenth century. The empress had asked Euler to use his mathematical chops to demonstrate the existence of God to a sceptical Diderot. The mathematician obliged, declaring in front of a packed crowd: 'Sir, $\frac{a+b^n}{n}$, hence God exists; reply!' Diderot, unacquainted with algebra, was left utterly perplexed. He duly took his leave, defeated and humiliated.

This anecdote is almost certainly a fabrication.[1] For one thing, Diderot was no mathematical slouch, and he would have seen through such sophistry. In any case, Euler, serious mathematician that he was, would never have resorted to such a trivial line of argument.

But the story perfectly captures the mystical aura that mathematics exudes. Many arguments, often made in tangential disciplines such as philosophy and science, invoke mathematical concepts and arguments that are difficult for laypeople to scrutinise, and this perversely becomes part of their appeal.

One ethical view that has spawned such arguments is 'long-termism': the notion that humanity should prioritise the needs of the distant future because it theoretically contains hugely more lives than the present; another is effective altruism, a utilitarian approach that seeks to maximise good for society. The arguments made by proponents of these ethical frameworks contain speculative, numerical wanderings that, upon close inspection, border on fantastical. Consider, for example, this claim by Oxford philosopher Nick Bostrom ('existential risk' here refers to threats that could wipe out all of humankind):

> [T]he expected value of reducing existential risk by a mere *one billionth of one billionth of one percentage point* is worth a hundred billion times as much as a billion human lives.[2]

Bostrom is drawing on the concept of an 'expected value', an idea from probability that gives us a way of assigning values to uncertain outcomes. Suppose you buy a scratchcard, knowing there is a 10 per cent chance of scooping a £50 cash prize and a 90 per cent chance of receiving nothing. The expected value in this case is a measure of your scratchcard's worth, and the way to calculate it is to multiply your probability of success, $\frac{1}{10}$, by the payout of £50, giving £5. That may seem strange because there are only two outcomes for a single bet (you win £50 or you leave with nothing). But if you bought a whole stash of scratchcards, then the expected value is a useful measure of how much you will win on average (the more scratchcards you buy, the more accurate this prediction will be).

We can apply the same concept in less obvious ways. At some point you may have been faced with the decision of whether to pay extra to insure your new phone against damage. If you judge that there's a 20 per cent (or one in five) chance of your phone being accidentally damaged within the period of cover, and the repairs would cost £200, then the expected value of the warranty cover is $\frac{1}{5} \times £200 = £40$. If the warranty costs more than this, you might be minded to take your chances.

This is essentially the mechanism at play in Bostrom's argument, although the numbers involved are more extreme (which, I suspect, is one reason they are not interrogated more deeply). Even with an example as straightforward as phone insurance, there is some degree of speculation – to produce an expected value, we have first to assign the one-in-five chance of the phone being accidentally damaged. Why not one in four? Bostrom's calculation is based on assumptions wilder than this and, for that matter, probably wilder than anything you can imagine. He estimates, for instance, that in the distant future (no specific timeframe is provided), with advances in fields such as AI, interstellar space travel and computerised brain simulations, humans have the potential to give rise to 10^{52} conscious beings (that may exist in some virtual simulation). He also assigns a 1 per cent chance of all this coming to pass, before feeding these numbers into the expected value calculation. If the numbers feel arbitrary that's because they are; they are certainly not grounded in any serious scientific theory.

There are further problems with Bostrom's approach. If you subscribe to the logic of an expected value for such extreme values, then, by the same token, you'll be persuaded to take all manner of implausible bets. If you receive an email from a Nigerian prince promising to leave you billions of dollars' worth of inheritance provided you deposit the comparatively small sum of £5, your prudent course of action would be to relegate the email to spam. Yet according to the expected value framework, if you think there's even a 0.000001 per cent chance of the offer being legitimate, the rational step is to take your chances, because the offer is 'worth' £10. I'm sure you'll agree there are better uses for your money.

What we're seeing here is something we already glimpsed in Chapter 5 – the limits of theories on human behaviour that reduce us to rational agents seeking to maximise utility. Expected value takes no account, for instance, of our aversion to loss. Many people justifiably resist the temptation of a £50 jackpot on a scratchcard because they cannot stomach the prospect of losing money (especially given the slim odds of success). A recurring theme of this book is that as much as mathematics enriches our view of the world, it cannot fully shape it. Every mental model brings its own assumptions, its own limits. Our foray into high-dimensional spaces seemed to offer a limitless way of thinking, but we also saw that many situations are best described with spaces that have a fractional dimension. A whole chapter was devoted to gradients, yet the smooth functions that those ideas apply to turn out to be vanishingly rare. Our attempt to formalise the logical certitude of mathematics came up short, first with the discovery that no collection of axioms can be considered perfect, then with the recognition that our notions of truth are fuzzier than the binary choice between true and false allows.

The misuse of expected value similarly reminds us of the limits of mathematised views of the world. In the most extreme version of expected value, where we countenance infinite pay-offs, we're left with such banalities as Pascal's wager, in which the French mathematician argued for believing in God as follows:

- God either exists or He does not.

- Assign a probability to God existing – you can make it however small you like, but it must be positive; there's *some chance* of it being true.
- Now consider your choices in each of the two scenarios:
  » If God exists: believing in God brings infinite reward (e.g. eternal salvation) but not believing in God results in infinite punishment (eternal damnation).
  » If God does not exist: believing in God now incurs finite loss (a mortal life spent entertaining false beliefs), whereas not believing in God brings finite reward (a life of enlightenment).

Your expected value for believing in God is therefore infinite, Pascal argues, because the infinite reward for being correct (however remote a possibility that might be) outweighs the finite cost of being wrong. The only thing Pascal's thought experiment demonstrates is to use infinity with an abundance of caution. This book has flirted with the infinite on more than one occasion – the infinitesimal units of analysis in calculus, Cantor's infinite tower of ever-greater infinities, the unrelenting recursion of fractals. Mathematicians have learned to incorporate infinity into meaningful contexts, but they also know when to keep it at arm's length.

The upshot of Bostrom's argument is that the expected number of lives saved with even a marginal reduction in existential risk (which amounts to 'a hundred billion times as much as a billion human lives') far exceeds the population of humans alive today. We could dismiss these kinds of arguments as playful thought experiments, but they are taken seriously in many quarters. The Future of Humanity Institute, founded by Bostrom and which has brought together philosophers, computer scientists, economists and, yes, mathematicians, has been a germinating ground for ideas on long-termism, framing much of the debate on existential risks associated with AI.[3]

When you combine the ideas of long-termism with the prospect of 10^{52} future lives, you can see why present-day concerns such as geopolitical turmoil and runaway climate change are deemed relatively unimportant (and why endeavours such as the development of AI, which is seen as the pathway to securing those future lives, is deemed a moral imperative by many who espouse the long-termist philosophy). Sure, current threats to humanity may impact the lives of billions today and in the near future, but that pales in comparison to the size of future civilisations. The undisciplined use of mathematical models gives no room, in this instance, to the myriad ethical concerns inherent in such judgements. Critics have also warned that this fixation occludes real, present-day problems associated with AI such as biased algorithms or the widening of socio-economic inequality due to automation.[4]

I think the ideas of long-termism are interesting, worthy of debate certainly. What I contest, though, is the way mathematics is weaponised to give their central claims clout. The battle for our climate, not to mention our democratic ideals, is already being lost to campaigns of disinformation and the distrust being seeded in institutions and honest-to-god scientific enterprise. Arguments that downplay present-day threats need to be called out for the unsubstantiated musings that they are, not least when they exploit maths to gain favour and credibility.

The mental models of mathematics are among the most powerful tools we have for understanding our world. They illuminate the patterns of our everyday lives, shape our arguments and expand our mental horizons. And they are available to every one of us. Reading this book, I hope the aura of mathematics has been lifted and that you feel a little closer to the priesthood of mathematicians. You have at your disposal an arsenal of thinking tools that can sharpen your sense of the world and help counter the false claims that pervade our everyday discourse, so many of which are rooted in bad applications of mathematical concepts. Misuse of maths, deliberate or otherwise, does the subject a disservice by breeding mistrust in logic and evidence. I hope you will instead use your mental models to inspire clarity of thought, in yourself and in others.

References

Introduction

1   See, for example, B. Russell, 'The Study of Mathematics', in *Mysticism and Logic and Other Essays* (Longman, 1919), G. H. Hardy, *A Mathematician's Apology* (Cambridge University Press, 1992) and P. Lockhart, *A Mathematician's Lament* (Bellevue Literary Press, 2009).

2   Attributed to Henry John Stephen Smith in H. Eves, *Mathematical Circles Squared* (Prindle, Weber and Schmidt, 1972).

3   A. Flexner, 'The Usefulness of Useless Knowledge', *Harpers*, issue 179 (1939), pp. 544–552.

Chapter 1 – The Continuum

1   M. Farisco et al., 'The Intrinsic Activity of the Brain and its Relation to Levels and Disorders of Consciousness', *Mind & Matter*, vol. 15(2) (2017), pp. 197–219.

2   S. Strogatz, *Infinite Powers* (Atlantic Books, 2019), p. xiv.

3   S. Ahearn, 'Tolstoy's Integration Metaphor from War and Peace', *American Mathematical Monthly*, vol. 112(7) (2005), pp. 631–638.

4   L. Tolstoy, *War and Peace* (trans. L. and A. Maude), G. Gibian, ed. (W. W. Norton, 1966), p. 918.

5   K. Devlin, 'How to Stabilize a Wobbly Table', *Keith Devlin blog* (February 2007). https://profkeithdevlin.org/devlins-angle/2007-posts/#feb07

6   B. Baritompa et al., 'Mathematical Table Turning Revisited', arXiv.org (19 November 2005). https://arxiv.org/abs/math/0511490

7   M. Gladwell, *Outliers* (Penguin, 2009).

8   'Children Do Better in School if They Were Born in the
     Autumn', Centre for Educational Neuroscience, accessed
     2 November 2024 at https://www.educationalneuroscience.org.
     uk/resources/neuromyth-or-neurofact/children-do-better-in-
     school-if-they-were-born-in-the-autumn/

Chapter 2 – Gradients

1   J. Clear, *Atomic Habits* (Random House Business, 2018),
     p. 15.
2   W. Malthus, *An Essay on the Principle of Population*
     (J. Johnson, 1798).
3   H. Rosling, *Factfulness* (Sceptre, 2018), pp. 84–87.
4   R. Kurzweil, 'The Law of Accelerating Returns', the
     Kurzweil Library + collections (1 February 2024), accessed
     14 September 2024 at https://www.writingsbyraykurzweil.com/
     the-law-of-accelerating-returns
5   F. Kafka, *The Diaries of Franz Kafka* (Schocken classics)
     (Deckled Edge Edition): *The Diaries 1910–1923* (The Schocken
     Kafka Library) (Schocken Books, 1988), p. 281.
6   H. Rossi, 'Mathematics is an Edifice, not a Toolbox', *Notices
     of the American Mathematical Society*, vol. 43(10) (1996).
     http://www.ams.org/notices/199610/page2.pdf

Chapter 3 – Relations

1   From George Orwell's review of Bertrand Russell's book
     Power: A New Social Analysis, published in *The Adelphi*
     magazine (January 1939).
2   From F. Dostoyevsky, *Notes from Underground* (East India
     Publishing Company, 2022).
3   K. Carr (@kareem_carr), 'I don't know who needs to hear this
     but if someone says "2 + 2 = 5", the correct response is "What
     are your definitions and axioms?" not a rant about the decline
     of Western civilization.' [Tweet], X (30 July 2020). https://x.
     com/kareem_carr/status/1288838380625821696
4   J. Lindsay, '2 + 2 Never Equals 5', *New Discourses* (3 August
     2020). https://newdiscourses.com/2020/08/2-plus-2-never-
     equals-5/

5   B. Berry, 'Why Was 5 × 3 = 5 + 5 + 5 Marked Wrong?', *Huffington Post* (4 November 2015). https://www.huffpost. com/entry/why-was-5-x-3-5-5-5-marked-wrong_b_8468722

6   K. Buzzard, 'Grothendeick's Use of Equality', *arXiv.org* (16 May 2024). https://arxiv.org/pdf/2405.10387

7   H. Tajfel H et al., 'Social Categorization and Intergroup Behavior', *European Journal of Social Psychology*, vol. 1(2) (1971), pp. 149–178.

8   N Richter et al., 'The Effects of Minimal Group Membership on Young Preschoolers' Social Preferences, Estimates of Similarity, and Behavioral Attribution', *Collabra*, vol. 2(1) (January 2016), pp. 1–8.

9   T. Pychyl, 'Intransitive Preference Structures: The Procrastination Trap', *Psychology Today* (17 April 2008). https://www.psychologytoday.com/gb/blog/dont-delay/200804/ intransitive-preference-structures-the-procrastination-trap

10 A. Tangian, 'Unlikelihood of Condorcet's Paradox in a Large Society', *Social Choice and Welfare*, vol. 17(2) (2000), pp. 337–365.

Chapter 4 – Sets

1   L. Dartnell, *The Knowledge* (Vintage, 2015).

2   K. Crenshaw, 'Demarginalizing the Intersection of Race and Sex: A Black Feminist Critique of Antidiscrimination Doctrine, Feminist Theory and Antiracist Policies', *University of Chicago Legal Forum*, vol. 1989(1) (1989). See also J. Coaston, 'The Intersectionality Wars', *Vox* (28 May 2019). https://www.vox.com/the-highlight/2019/5/20/18542843/ intersectionality-conservatism-law-race-gender-discrimination

3   M. Hall, 'On the Cover: Vhat? Vhere? Venn', *Chalkdust* magazine (22 November 2021). https://chalkdustmagazine. com/features/on-the-cover-vhat-vhere-venn/

4   J. Aron, 'Discover the Beauty of Extreme Venn Diagrams', *New Scientist* (13 August 2012). https://www.newscientist. com/gallery/venn/

5   D. Bennett, 'Origins of the Venn Diagram', in M. Zack and E. Landry (eds), *Research in History and Philosophy of*

Mathematics. Proceedings of the Canadian Society for History and Philosophy of Mathematics (Birkhäuser, 2015).

6   I. Spence, 'No Humble Pie: The Origins and Usage of a Statistical Chart', *Journal of Educational and Behavioural Statistics*, vol. 30(4) (December 2005). https://doi.org/10.3102/10769986030004353

7   B. Minto, 'MECE: I Invented It, So I Get to Say How to Pronounce It', McKinsey Alumni Center, accessed 12 August 2024 at https://www.mckinsey.com/alumni/news-and-events/global-news/alumni-news/barbara-minto-mece-i-invented-it-so-i-get-to-say-how-to-pronounce-it

8   S. McChrystal, *Team of Teams* (Penguin, 2015).

Chapter 5 – Axioms

1   F. O'Sullivan and D. Zuidijk, 'The 15-Minute City Freakout is a Case Study in Conspiracy Paranoia', *Bloomberg UK* (2 March 2023). https://www.bloomberg.com/news/articles/2023-03-02/how-did-the-15-minute-city-get-tangled-up-in-a-far-right-conspiracy

2   D. Freeman et al., 'Coronavirus Conspiracy Beliefs, Mistrust, and Compliance with Government Guidelines in England', *Psychological Medicine*, vol. 52(2) (2022), pp. 251–263.

3   K. Popper, *The Open Society and its Enemies* (Princeton University Press, 2013), ch. 7.

4   The discussion here centres on the seminal work of John von Neumann and Oskar Morgenstern – see J. von Neumann and O. Morgenstern, *Theory of Games and Economic Behaviour* (Princeton University Press, 1953).

5   J. Muller, *The Tyranny of Metrics* (Princeton University Press, 2019).

Chapter 6 – Logic

1   C. Wason and P. Johnson-Laird, *Psychology of Reasoning: Structure and Content* (Harvard University Press, 1972).

Chapter 7 – Combinatories

1   S. Veiga et al., 'How Mixed Relay Teams in Swimming Should Be Organized for International Championship Success', *Frontiers in Psychology*, vol. 12 (2021). https://doi.org/10.3389/fpsyg.2021.573285

2   D. Messick and R. Kramer (eds), *The Psychology of Leadership* (Routledge, 2004), ch. 6.

3   C. Myers, 'How to Lead When Your Business Leans Towards Chaos', *Forbes* (19 September 2017), accessed 2 November 2024 at https://www.forbes.com/sites/chrismyers/2017/09/19/entrepreneurial-entropy-how-to-lead-when-your-business-leans-toward-chaos/

4   E. McIrvine and M. Tribus, 'Energy and Information', *Scientific American*, vol. 225(3) (1971), pp. 179–190.

5   C. Shannon, 'Programming a Computer for Playing Chess', *Philosophical Magazine,* vol. 41(314) (March 1950).

Chapter 8 – Dimensionality

1   A. Heywood, *Political Ideologies: An Introduction* (6th edn) (Macmillan International Higher Education, 2017).

2   C. Alós-Ferrer and D-G Grranić, 'Political Space Representations with Approval Data', *Electoral Studies*, vol. 29 (September 2015), pp. 56–71.

3   See, for example, A. Reddy, 'The Eugenic Origins of IQ Testing: Implications for Post-Atkins Litigation', *DePaul Law Review*, vol. 57(3) (2008), pp. 667–677 and P. Levine, *Eugenics: A Very Short Introduction* (Oxford Academic, 2017), ch. 2.

4   J. Flynn, *Are We Getting Smarter?* (Cambridge University Press, 2012).

5   H. Gardner, *Frames of Mind* (Basic Books, 1993).

6   M. Mitchell, 'Debates on the Nature of Artificial General Intelligence', *Science*, vol. 383(6689) (2024). https://www.science.org/doi/10.1126/science.ado7069

7   D. Goel and D. Batra, 'Predicting User Preference for Movies Using NetFlix Database', Department of Electrical and

Computer Engineering, Carnegie Mellon University (2006), accessed 7 June 2024 at https://www.cs.cmu.edu/~epxing/Class/10701-06f/project-reports/goel_batra.pdf

8   The official test can be taken at https://www.mbtionline.com/

9   C. Campbell et al. (eds), *Problems of Personality: Studies Presented to Dr Morton Prince, Pioneer in American Psychopathology* (1st edn) (Routledge, 1925).

10   L. Stricker and J. Ross, 'An Assessment of Some Structural Properties of the Jungian Personality Typology', *Journal of Abnormal & Social Psychology*, vol. 68(1) (January 1964), pp. 62–71.

11   T. Judge et al., 'Five-Factor Model of Personality and Job Satisfaction: A Meta-analysis', *Journal of Applied Psychology*, 87(3) (2002), pp. 530–541.

12   S. Seibertand and M. Kraimer, 'The Five-Factor Model of Personality and Career Success', *Journal of Vocational Behavior*, vol. 58 (2001), pp. 1–21.

Chapter 9 – Distance

1   N. Mutebi and A. Hobbs, 'The Impact of Remote and Hybrid Working on Workers and Organisations', UK Parliament Research Briefing (17 October 2022), accessed 14 June 2024 at https://post.parliament.uk/research-briefings/post-pb-0049/#:~:text=there%20are%20limited%20data%20on,working%20due%20to%20the%20pandemic

2   See, for example, C. O'Neil, *Weapons of Math Destruction* (Penguin, 2017).

3   B. Stevenson, *Just Mercy* (Scribe, 2015), p. 12.

4   D. Bar-Tal, 'Causes and Consequences of Delegitimization: Models of Conflict and Ethnocentrism', *Journal of Social Issues*, vol. 46(1) (1990), pp. 65–81.

5   B. Leidner et al., 'Ingroup Glorification, Moral Disengagement, and Justice in the Context of Collective Violence', *Personality and Social Psychology Bulletin*, vol. 36 (2010), pp. 1115–1129.

Chapter 10 – Fractals

1 B. Mandelbrot, *The Fractal Geometry of Nature* (W. H. Freeman, 1982).

2 B. Mandelbrot, 'How Long Is the Coastline of Britain?', *Science, New Series,* vol. 156(3775) (May 1967), pp. 636–638. http://li.mit.edu/Stuff/CNSE/Paper/Mandelbrot67Science.pdf

3 A. Bhattacharya et al., 'Building Fractal Advantage in a Fragmenting World', Boston Consulting Group (15 November 2021), accessed 18 October 2024 at https://www.bcg.com/publications/2021/fractal-advantage-in-a-fragmenting-world

4 B. Hakim, 'Reviving the Rule System', *Cities*, vol. 18(2) (2001), pp. 87–92. https://www.archnet.org/publications/10461

5 J. Cepelewicz, 'The Quest to Decode the Mandelbrot Set, Math's Famed Fractal', *Quanta Magazine* (26 January 2024). https://www.quantamagazine.org/the-quest-to-decode-the-mandelbrot-set-maths-famed-fractal-20240126/

6 This example is taken from T. Pratchett and N. Gaiman, *Good Omens* (Gollancz, 2015).

7 From M. Lewis, *The Undoing Project* (Penguin, 2017).

8 B. Klaas, *Fluke* (John Murray, 2024).

9 D. Hoffman, *The Case Against Reality* (Penguin, 2020).

Epilogue

1 See D. Chene, 'On Bad Anecdotes and Good Fun', *New APPS blog* (2012), accessed 10 November 2024 at https://www.newappsblog.com/2012/07/on-bad-anecdotes-and-good-fun.html

2 N. Bostrom, 'Existential Risk as Global Priority', *Global Policy*, vol. 4(1) (2013), accessed 12 October 2024 at https://existential-risk.com/concept

3 J. Broughel, 'Effective Altruism Contributed to the Fiasco at OpenAI', *Forbes* (2023), accessed 10 December 2024 at https://www.forbes.com/sites/jamesbroughel/2023/11/20/effective-altruism-contributed-to-the-fiasco-at-openai/

4 For critiques of long-termism, see, for instance, A. Kaspersen and W. Wallach, 'Long-termism: An Ethical Trojan Horse', Carnegie Council for Ethics in International Affairs (2022),

accessed 30 November 2024 at https://www.carnegiecouncil. org/media/article/long-termism-ethical-trojan-horse; and É. Torres, 'Why Effective Altruism and "Longtermism" Are Toxic Ideologies', *Current Affairs* (2023), accessed 15 November 2024 at https://www.currentaffairs.org/news/2023/05/why-effective-altruism-and-longtermism-are-toxic-ideologies

Acknowledgements

It is often said that to attempt a second marathon, enough time needs to pass for a runner to forget the endeavour of the first. I know this all too well because fifteen years have passed since my first marathon and I've yet to make it to another.

If the same is true of writing books, then I am indebted to the many people who have helped me avoid the same fate and get over the finish line of my second outing.

My agent Doug Young was the first to champion my writing and I am grateful for his unwavering support and advocacy, and for never failing to offer a reassuring word during periods of doubt.

It was my editor, Nick Humphrey, who persuaded me that this book could be more than a long list of mathematical tips and tricks and I am grateful for the loftier ambitions he pushed on me (and for his meticulous line edits). My thanks also go to Nick Allen, Georgina Difford and the whole editorial team for whipping this book into shape. A special thanks must be reserved for Helen Conford who backed this project when it was little more than sketches on a mental napkin.

A small army of mathematicians vetted the technical details in the book and supplied me with no shortage of suggestions for how to connect abstract topics with real-world thinking. My thanks in particular to Ed Border, Taimur Abdaal, Steve Buckley, Victor Flynn, Danny Kodicek, Alan Lauder, Ayliean Macdonald, Sophie Maclean and James Tanton. Any mathematical errors are surely mine. A further thanks to Michael Marshall for his insights into how conspiracy theories take flight.

The two mathematicians I must thank above all others are Hilary Priestley and Graham Nelson, my tutors at Oxford who, two decades ago, planted the seeds for this book by instilling in me the deepest appreciation for pure mathematics. They taught me what it means to think like a mathematician.

During the course of writing this book I have had the privilege of establishing the Parallel Academy, an online maths school of excellence. I am grateful to Parallel's founder Simon Singh not only for his popular maths books that inspired me a generation ago, but for the opportunity to make a living from a genuine passion project. I am equally grateful to the thousands of students I have interacted with over the past few years. To see these eager young minds thinking like the mathematicians they aspire is a constant source of inspiration.

My biggest thanks go to my family, for keeping me rooted during the past few years. My wife Kawther has had to endure the full spectrum of my emotions throughout this project. She has served as therapist, coach and editor, all at once (to all aspiring authors: it really helps if your partner is a far superior writer to you). My children Leena and Elias, six and four (*shopping in the store!*) at the time of this book's publication, have provided much needed relief in between writing sessions. There is no greater privilege than dedicating a work to a loved one, and no choice more obvious for this book. Leena got the last one so this one's for you, Elias.

Index

Page references in *italics* indicate images.